Gaia and Climate Change

'A magnificent gift: Primavesi blends action-provoking clarity with biblical hermeneutics and Continental philosophy to bring Gaia into theological focus. Taking on the momentous complexity of the climate crisis, she does not yield to the temptations of eco-apocalypse. With inspiring composure she makes a profound contribution to tehomic theology.' – Catherine Keller, Professor of Constructive Theology, *Drew University*

'With gentle yet persistent skill Primavesi credibly links climate change, the perilous journey of western Christianity and a radical recovery of Gospel Kingdom. Her book offers crucial insight and time-tested passion to assist theology in what may be its most defining challenge today.' – Stephen Dunn, founder of the *Elliott Allen institute for Theology and Ecology*

James Lovelock's Gaia theory revolutionized the understanding of our place and role in the global environment. It is now accepted that our activities over the past two hundred years have contributed to and accelerated the extreme weather events associated with climate change. The fact that those activities materialized, for the most part, from within Western Christian communities makes it imperative to assess and to change their theological climate: one characterized by routine use of violent, imperialist images of God.

The basis for change explored here is that of gift events, particularly as evidenced in Jesus' life and sayings. Its legacy of love of enemies and forgiveness offers a basis for nonviolent theological and practical approaches to our situatedness within the community of life. These are also Gaian responses, as they include foregoing a perception of ourselves as belonging to an elect group given power by God over earth's life-support systems and over all those dependent on them, whether human or more-than-human. The degree to which we change this self-perception will determine how we affect, for good or ill, not only the givenness of the climate in future but the givenness of all future life on earth.

Anne Primavesi is presently a Fellow of the Westar Institute and the Jesus Seminar, Santa Rosa, California and author of *Sacred Gaia* (Routledge, 2000) and Gaia's Gift (Routledge, 2003).

Gaia and Climate Change

A theology of gift events

Anne Primavesi

R Routledge
Taylor & Francis Group

LONDON AND NEW YORK

First published 2009 by Routledge
2 Park Square, Milton Park, Abingdon, Oxon OX14 4RN

Simultaneously published in the USA and Canada
by Routledge
270 Madison Ave., New York, NY 100016

Routledge is an imprint of the Taylor & Francis Group

Typeset in Galliard by
GreenGate Publishing Services, Tonbridge, Kent

Printed and bound in Great Britain by TJ International Ltd,
Padstow, Cornwall

British Library Cataloguing in Publication Data
A catalogue record for this book is available from the British Library

Library of Congress Cataloging in Publication Data
Primavesi, Anne, 1934-
Gaia and climate change : a theology of gift events / Anne Primavesi.
p. cm.
Includes bibliographical references (p.) and index.
1. Human ecology--Religious aspects--Christianity. 2. Climatic changes--
Effect of human beings on. 3. Climatic changes--Moral and ethical aspects.
4. Gaia hypothesis. I. Title.
BT695.5.P745 2008
261.8'8--dc22
2008007858

ISBN 10: 0-415-47157-5 (hbk)
ISBN 10: 0-415-47158-3 (pbk)
ISBN 10: 0-203-89171-6 (ebk)

ISBN 13: 978–0-415–47157–2 (hbk)
ISBN 13: 978–0-415–47158–9 (pbk)
ISBN 13: 978–0-203–89171–1 (ebk)

In memory of Robert Funk
and with gratitude for the scholarship and inspiration of the Fellows of the Westar Institute and the Jesus Seminar

Contents

Foreword   ix
Acknowledgements   xi
Introduction   1

1 The Context of Climate Change   9

2 The Seminal Event   17

3 The First Historic Event   28

4 The Second Historic Event   38

5 The Third Historic Event   55

6 The Givenness of Events   66

7 The Economy of Gift Events   80

8 Changing God's Image   89

9 The Gift Event of Jesus   100

10 The God of Jesus – or of Caesar?   108

11 What Jesus Said   118

12 Beginning Something New   130

Bibliography   141
Index   146

Foreword

Anne Primavesi introduces her fascinating book by pointing out how fast, in recent times, our traditional patchy, atomized notions about climate change have been giving way to more comprehensive, Gaian ones, and she reasonably comments that this change in our global perspective 'calls for a corresponding change in perspective for those of us who are theologians'. So it does. But I would like to suggest here that perhaps more of us should count ourselves as theologians in this sense than might at first be inclined to do so.

All of us, including those who might claim to have no theology at all, still have mental maps which do the same work for us that doctrines about God and gods did in the past. They express strong, dramatic visions of how the universe essentially works and especially of our own place in it – visions which determine not just how we think, but also how we feel and act.

Of course today these visions are supposed to be quite impersonal and scientific, based on an ascetic rejection of all 'grand narratives'. And they do indeed incorporate material drawn from the physical sciences. But we humans cannot live emotionally on bare physical facts any more than we can digest stones. We are natural dramatizers, always apt to find patterns that will give us the meaning that we need, and when we drop one such pattern of relation, we tend to replace it with another. Thus, since Western culture ceased to talk easily in Christian terms, it has generated a rich mix of beliefs centring on the heroic role of the human race itself – stories dramatizing various notions of our progress and evolution, myths of fierce competition between species, tales of a cosmic casino powered by natural selection, fantasies of transformation and mechanization and sagas of dealings with aliens. In all these dramas the central character is, of course, no longer God but essentially ourselves. Theism gives way to Humanism.

What has struck me particularly in reading this book is how often this new central character, this new visionary focus, seems merely to reproduce the notorious faults of his predecessor. As the author outlines the changes that are now needed, she diagnoses the blots that have marred the traditional notion of God, particularly the constant over-emphasis on his unaccountable power, arbitrary will and consequent violence. She remarks that the whole traditional notion of divine creation has, in Catherine Keller's words, been 'barnacled with stereotypes of a great supernatural surge of unlocatable father-power'. To illustrate its political effects she quotes from a speech made by a senator before the US Senate in 1900 – 'God has made us the master organizers of the world to establish system where chaos reigns. He [*sic*] has made us adept in government that we may administer government among savages and senile peoples.'

As Anne Primavesi points out, these patriarchal legacies have led to a kind of overconfidence which unites in power the traditional God and his worshippers. It gives a sense of being necessarily in charge and thus generates the anthropocentric attitudes which now prevent Westerners from understanding that they really are threatened by climate change. Again and again, as she examines particular issues on which a Gaian attitude requires different conduct, this strange notion of humanity as somehow separate and immune from planetary processes emerges as what blocks understanding. It has deep roots in our theological traditions – roots which are still active despite the widespread current conviction that this powerful God is dead. As she says, 'We must confront the role that Christian violence-of-God traditions have played in causing the problems raised by climate change and in justifying our part in them. Awareness of, and attention to, this state of affairs is a prerequisite for remedying it.'

This topic is, of course, only one small strand in the very complex pattern of connections which she traces as she sorts out the various elements in the tradition and asks what we ought now to keep and what we should alter. Most of this admirable book is actually occupied by a fascinating investigation of the concept of gift – of how we should work out a new sense of life as freely given to us, without needing to dramatize the giver as arbitrary and alien. On this topic she draws freely on contemporary biblical scholarship into the historical Jesus to expose defects in the traditional Christian image of God. In doing this, however, she does naturally lay some emphasis on those defects. And here it strikes me as especially interesting that something so apparently modern as our strange current confidence in our own capacity to run the universe efficiently should draw strength from a religious tradition which many people now believe to be obsolete. This does, perhaps, suggest, that theology is not just a private concern for professional theologians – that the patterns on which earlier people have supposed the cosmos to be organized have not just been idle superstitions, but serious and influential elements in the development of their thought. Perhaps, as Anne Primavesi suggests, they are something that the rest of us might do well to look at too.

Mary Midgley

Acknowledgements

In June 2004 James and Sandy Lovelock convened a Conference on Climate Change at Dartington Hall and included me in the proceedings as one of a group of non-scientists concerned about the issues it raises. This was yet another gift to me from them, one that has deeply affected my work from then on. I also found a real sense of commitment to meaningful change in religious attitudes to such problems during the aptly titled 'Ground for Hope: Faith, Justice and the Earth', a Transdisciplinary Colloquium held at Drew University in Autumn 2005. The proceedings have now been published by Fordham University Press as *Ecospirit: Religions and Philosophies for the Earth,* edited by Laurel Kearns and Catherine Keller. Then in Spring 2006 I participated in the 11th Conference of the European Society for the Study of Science and Theology in Iasi, Romania, entitled 'Sustaining Diversity: Science, Theology and the Futures of Creation'. Last, but by no means least, in May 2007 I was privileged to take part in the opening conference on 'Nature, Space and the Sacred' in Bamberg, Germany, organized by the European Forum for the Study of Religion and the Environment.

These meetings and the dialogues they engendered have given impetus, content and direction to this book. On a smaller but no less important scale has been the opportunity to share concerned responses to climate change with groups and individuals, particularly at the meetings of the Gaianet group in London. In Ireland the religious communities at Pairc an Tobair and Bru na Cruinne have inspired me with their lifestyle and dedication to the needs of the earth. Lucy Mooney has given hospitality and support in abundance. Last, but by no means least, the Edward and Celia James Book Fund has enabled my immediate access to essential reading matter.

There are individuals, however, who are owed a particular debt of gratitude not only by me but by all who read this book. Barbara Turner, Senior Research Scientist Emerita in Biological Sciences at Stanford University, has read, checked and commented on the whole text, thereby saving me from displaying ignorance and my scientist friends from apoplexy. My editor, Lesley Riddle, made an act of faith in commissioning what was, at the time, no more than a vague project and then seeing it through to completion. Glynn Gorick made that completion visible by providing a third beautiful and thought-provoking cover for this 'Gaia' book also.

My husband Mark has guided the whole process, and me, through every stage from start to finish.

Love's mysteries in souls do grow. And yet the body is its book.

Introduction

In 1998, I suggested in an essay entitled 'The Recovery of Wisdom: Gaia Theory and Environmental Policy' that the change in vision brought about through an engagement with James Lovelock's Gaia theory marked a phase change in human understanding of the environment: an understanding essential for the formulation of good environmental policies. The theory clarified why such policies had to be formulated and implemented within a context of the well-being of a larger whole, one that ultimately encompasses the global environment. Gaia theory does this, I argued, by providing us with a comprehensive view of diverse life support systems within the context of earth as a single system (Primavesi 1998: 75).

Ten years on, it is a 'given' that our localized climate systems can be properly understood only in terms of a global one. This in turn has led to a multidisciplinary expansion of Gaia theory in which viewing the planet from the perspective of geological time rather than that of human history has changed and deepened our scientific understanding of it and of our place within it.

My contention throughout has been, and remains, that this change in our global perspective calls for a corresponding change in perspective for those of us who are theologians. We must come to view our relationship with God in the light of earth's history and our role in its later stages. At a time when certain events – such as the increase in catastrophic weather conditions – are understood as emerging from global climate change, it has become a driving force within a significant sector of world politics, economics and social justice movements. It is now increasingly urgent to assess its impact on theological perspectives also.

The nature of events

It is this state of affairs that makes an exploration of what Caputo calls 'a theology of the event' worthwhile. It is, he says, a theology with an ancient pedigree, one that goes back to 'a famous narrative about a very seminal event indeed'. For, if we look at what happened in the first creation story in Genesis:

It was as if the great elements – the womblike deep (*tehom*), the formless void (*tohu wa bohu*), and the wind (*ruach*) were sleeping, and ... what Elohim did was to release the events that stirred within these great sleeping elements.

(Caputo 2007: 49–50)

Studying that story from the perspective of geological time we know that that seminal event of creation and those following on from it preceded any human witness to them or detailed knowledge of them. That fact is also part of the biblical record. God reminds Job:

Where were you when I laid the foundation of the earth? ...
Who determined its measurements – surely you know!
Or who stretched the line upon it?
On what were its bases sunk, or who laid its cornerstone?

(Job 38: 4–6)

Such a text alludes to and at the same time acknowledges our absence from the seminal events that formed the earth, its crust and its atmosphere. However, present-day advances in knowledge allow us to deduce them and to date them (in human terms) with some accuracy. This is also true of the events that followed them associated with the evolution of living organisms and their role in shaping the single dynamic geophysiological process Lovelock named Gaia. That continuing process – whose scope will always lie beyond our power to define, as it did for Job – also includes what are now named 'extinction events' such as that of the dinosaurs.

Out of that, however, emerged new mammalian life forms, including our own. As Caputo says:

The crucial move here lies in treating the event as something that is going on in words and things, as a potency that stirs within them and makes them restless with the event.

(Caputo 2007: 50)

The potency *within* the words used to describe an event is caught in Deleuze's image of an event being 'like a crystal' that becomes and grows 'only out of the edges or on the edge' (Deleuze 1990: 12). In other words, the limits of the language used to name or to describe an event are continually being transcended over time both in words and in what is actually happening. The potency within an event means that, in relation to the last extinction, we can see ourselves as its offspring.

This ability to see beyond limits and their effects, thereby transcending time, can be expressed theologically as re-envisioning events within a sacred domain: one that goes beyond every horizon set by space and time or by human standards. One theological name for this is the kingdom of God. A scientific one is *Sacred Gaia*. Both names express the insight that what happens in an event is not reducible to what we can measure, describe or define at any one time. There is an excess, an overflow beyond the reach of words or of symbolic representation. It is also the case, as we shall see in Hannah Arendt's discussion of three major events in human history, that from our perspective such events bear within them a potential for triumph or failure, for hope or despair – or rather, for both at once. For their impact cannot be restricted to any particular reaction to them, assessment of them or personal experience.

A potential for failure appears most obvious when looking at what scientists call 'extinction events', the significant and unpredictable loss of species chronicled in the fossil record. They make a quantitative distinction between phases in the patterns of these events. In background extinctions, species disappear at a low rate. In mass extinctions, as the name implies, there are high rates of disappearance in which formerly successful species or groups of species meet abrupt ends presumably because they could neither anticipate nor adapt to the agents bringing about change.

The fifth and best known mass extinction, that of the dinosaurs some 65 million years ago, restructured the biosphere with an equally unpredictable set of survivors, placental mammals. The subsequent 'biotic rebound' from this event allowed those mammals to evolve eventually into the human beings we are today. It remains unpredictable how and when the growing pattern of background extinctions may increase with global climate change; or how we in turn will use our potential as a species for the good or ill of those who come after us (Leakey and Lewin 1996: 62–8).

Theological adaptation in a time of climate change

An allied question in a time of climate change is what is happening to our potential for expressing a relationship with God? In Caputo's 'theology of the event' what happens to us is the *event* harboured in the name of God: the vocative, evocative, provocative force sheltered by that name (Caputo 2006: 13). This gnomic statement will be looked at from different perspectives throughout this book. Here it can be said that the potency that stirs within this name emerges for me, to a large extent, from what 'went on' in the life of Jesus. Even there, however, or perhaps especially there, it can never be reduced to any human expression, definition or representation. Neither is it reducible to any one happening nor can it be contracted into some finite form.

Nevertheless, envisioning what was 'released' into view in Jesus' life of the event harboured in the name of God allows us to explore the religious resources implicit in this name and be nourished by their force (Caputo 2007: 50–1). This exploration will be done without any desire to claim or even to imply that everything harboured in the name of God was revealed in Jesus – that would contradict the very notion of event employed here – or that other religions are any more or less inadequate or unnecessary in this regard.

At the same time, the oscillation between positive and negative meanings discerned in events discloses the mysterious nature of the relationships 'going on', not only as they did in the life of Jesus but also within normal, everyday events today. I see them as elements within what the poet Brendan Kennelly calls 'the mystery of giving'. In company with philosophers like Derrida, I shall call these elements 'gift events' characterized by the actual relationships of receiving and giving within and between the members of the community of life on earth, together with what we assess as their positive and negative effects.

> What have you that you did not receive? If then you received it, why do you boast as if it were not a gift?
>
> (1 Corinthians 4: 7)

Against this background I shall examine what happens in the event that, for Christians, is harboured in the name of God: the gift of Jesus. That theological exploration will be done in the context of the ongoing event we name 'climate change'. This means questioning and challenging traditional perceptions of our relationships with earth and, by extension, our relationships with each other and with God. This challenges us to see them as creating a single theological climate in which how we relate to one feeds back into our relationships with the others.

Overall, this integration of relationships will open up a nonviolent, non-consumerist, generous, compassionate and inclusive theological vision that can be seen as emergent in the life of Jesus. It presumes and evokes a lifestyle deeply committed to, and consistently involved in, relational events and processes where the freedom and unconditionality of our connectedness as receivers and givers marks us as his followers, whether or not we name ourselves Christians. Such a manner of discipleship necessarily releases us from an accepted image of God as an all-powerful, punitive, disembodied and yet implicitly violent force in all earthly lives.

Seeing our lives in terms of gift events constitutive of our relationship with God means accepting that they are 'beyond definition'. Yet in this process we also become more and more aware that these events make possible the lives and deaths of every organism and species – including our own. Taken as a whole they form a sacred domain that is not opposed to, but makes an unconditional claim on the kingdoms of this world and their short-term economic objectives. It challenges them by offering us a paradigm of what Catherine Keller calls a 'counter-imperial ecology of love'. At a time when climate change is heralded by an increase in extreme weather events that impact disproportionately on the lives of the poorest and therefore most vulnerable within society, she rightly sees that contemporary society is characterized by 'Our Mutually Assured Vulnerability' (Keller 2005: 131–3).

Such theologizing may appear too radical for those whose religious perspective is confined to a view of events within life on earth marked by supernatural constancy:

For everything there is a season, and a time for every matter under heaven:
a time to be born, and a time to die;
a time to plant, and a time to pluck up what is planted;
a time to kill, and a time to heal;
a time to break down, and a time to build up.

(Ecclesiates 3: 1–3 RSV)

Scientific adaptation to climate change

But it is no longer possible for us to interpret such religious expressions of faith in an irresponsible fashion: one that allows us to avoid responsibility for climate change. For this is a time when scientists tell us that an increase in catastrophic weather events, pointing to a 'critical threshold' within the global climate system, is due to a continuing rise in our carbon emissions into the atmosphere. If we cross that threshold, we go beyond a point where human intervention can stabilize the system. We will already have set in motion an irreversible major extinction event comparable to the other five extinction crises that the earth has previously experienced.

Setting out in detail the scientific evidence for this prognosis based on an understanding of positive feedback mechanisms, a Report from David Wasdell, Director of the Apollo-Gaia Project, to a meeting of policy makers in Madrid, 23 September 2007, in preparation for the Intergovernmental meeting on Climate Change in Bali in December 2007 concludes:

> We are now in the early stages of runaway climate change. There does not appear to be any naturally occurring negative feedback process available to contain its effects. Strategically, we have to generate a negative feedback intervention of sufficient power to overcome the now active positive amplifying feedback process. We then have to maintain its effectiveness during the remaining period of rising temperature, while temperature-driven positive feedbacks continue to operate. That is an extraordinarily difficult task, out of all comparison with strategies currently in place.
>
> (http://www.meridian.org.uk)

Faced with our mutually assured vulnerability, our contribution to it and the need to increase the potential of earth's life support systems for nourishing life, a provisionally articulated theological response might run as follows:

> In this time of climate change
> To everything there is just one season;
> The time is past when atmospheric CO_2 levels do not rise;
> For there is no time when our activities do not contribute to a rise in global temperatures.
> It is a time to plant trees, not to cut down forests;

A time to walk lightly on the earth, not to drive;
A time to cherish species, not to kill them;
A time to build up life support systems, and not needlessly to consume or waste them.

The harsh reality of climate change challenges us to integrate what we understand of past events that shaped the planet's life-sustaining processes with an understanding of their present critical state, and with a heightened awareness of the role we play in the further stabilization or destabilization of those processes. We know that a deceleration of climate change would eventually benefit the lives of those who come after us, particularly the world's poorest people. This deceleration must inevitably entail a lowering of consumption in the rich countries.

But because the immediate impact of lowering consumption affects the global economic system – built on continuing consumption – some argue that this negative effect of lowered consumption outweighs any positive effect in dealing constructively with climate change. Therefore such an impact needs to be seen ecologically in terms other than the growth of the fiscal economy, for example in the fact that the resources needed to sustain that growth are limited. It can be seen theologically in the context of our responsibility for 'the least of our brethren'. For what we do routinely and unthinkingly in the pursuit of our business interests has profound effects on the quality of their lives. Overall, reducing consumption involves fostering a deeper understanding of the inescapable interdependence between our lifestyles and the lives of all other members of the community of life on earth.

The givenness of events

It should be clear from what has already been said that naming an event at a particular moment in no way confines its momentum and effects to that point in time. On closer examination, every event discloses a hinterland whose dimensions we glimpse only with hindsight. That 'givenness' makes their present effects, to a certain extent, unpredictable. Increasing our sensitivity to the original, originating conditions of an event can, however, help us cope with its present effects. A current example is how sensitive we have become to the effects of overuse of carbon-based energy generation and the consequent search for alternatives.

Therefore, the theology outlined in this book will concentrate on two aspects of events: first, their nature as defined by philosophers, historians and theologians and second, their givenness. This latter aspect means, for science as well as theology, that searching for the origins and causes of events, such as climate change today, takes us back in time: back to a 'pre-original' elemental state where the given conditions for those events to emerge were created. In both theology and in science that elemental state is called 'chaos'.

In the Genesis text (and in other Near Eastern religious texts also) chaos is described as a set of relationships between elements: *tehom* (the deep), *tohu* (formlessness), *wa* (and), *bohu* (emptiness) and *ruach* (spirit). A certain order of co-existence emerges from these relationships both in the planet's topography and in the evolution of its inhabitants (Niditch 1985: 11–12, 18). In science, chaos is the study of disorder in the atmosphere, in the turbulent sea, in the fluctuations of wildlife populations, in the oscillations of the heart and brain. This appears as the irregular, discontinuous, erratic and disordered side of nature. Yet when scientists from different disciplines have sought connections between the different kinds of irregularity, physiologists, for example, have found a surprising non-linear order in the chaos that develops in the human heart and is a prime cause of sudden, unexplained death (Gleick 1998: 3–5).

Chaotic behaviour calls for a science of process rather than stasis, of becoming rather than being. It allows, indeed requires scientists to describe a given situation as one in which certain elements coalesce in unpredictable ways with unforeseen consequences. The connection between this and our ability to predict changes in climate is obvious. Meteorologist Edward Lorenz's analysis of weather systems focused on their being *sensitively dependent on initial conditions* as giving an acceptable definition of chaos. At the same time he stressed that those initial conditions (as discerned by us at a particular time and place) need not be the ones that existed when a system was created. They may also be the ones at the beginning of any stretch of time that interests an investigator, so that one person's initial conditions may be another's midstream or final conditions (Lorenz 1995: 8–9, 182).

In their book *Order out of Chaos* scientists Ilya Prigogine and Isabelle Stengers are conscious that the sensitivity of systems to fluctuations extends to societies. This sensitivity, they say, leads both to hope and to a threat: to hope, since even small fluctuations may grow and change the overall structure. Therefore, individual activity is not doomed to insignificance. On the other hand, this is also a threat, since the security of stable, permanent rules seems gone forever. This situation cannot, they say, inspire blind confidence. But it can, perhaps, inspire the same feeling of qualified hope that some Talmudic texts appear to have attributed to the God of Genesis.

> This God exclaims: 'Let's hope it works!' when, after twenty-six failed attempts, the present world emerges out of the chaotic heart of the preceding debris. This qualified hope has accompanied all the subsequent coevolutionary history of the world, a history 'branded with the mark of radical uncertainty'.
>
> (Prigogine and Stengers 1985: 313)

Such a qualified hope must accompany our efforts in dealing with the effects of climate change and inspire us at a time when theologians may not, so to speak, indeed must not pass the buck to an all-powerful God. That option is both proposed and rejected in a story from the Sufi tradition that describes a man going through the world seeing evil, suffering, sickness, death and war. Finally he rounds on God, demanding: 'God, why did you not *do something* about this?' God replies: 'I did do something about it. I made you.'

1 The Context of Climate Change

The 'inconvenient truth' of the troubling realities of climate change ought to alert theologians to the 'inconvenient truth' that certain readings of sacred texts, and traditional images based on them, have both provided and sanctioned images of God that have in turn sanctioned the violence of Christians.

(Nelson-Pallmeyer 2007: 7)

The present context

Climate change is the term used by scientists to describe the above-normal variations in present global weather patterns and temperature ranges caused by human activity. These variations and their effects are manifest in a measured increase in carbon emissions, the rate of ice cap melting, growing quantities of toxic waste, desertification, loss of biodiversity and alarming deficits in life support resource systems such as fresh water supplies. But the problems discerned by scientists are not theirs to solve alone: climate change is too important, too multi-causal for its effects to be dealt with by any one body of people or interests. Neither can the issues it raises be properly attended to by ignoring the role science has played in creating them; nor by simply tinkering with existing industrial technologies, economic systems or political strategies. Paraphrasing Einstein, we shall not solve climate change problems using the same sort of thinking that caused them in the first place.

There are various approaches to solving these problems being worked out, written about and implemented at present, many of them in response to an increasing number of authoritative reports that can be found on the Internet. These include the ones compiled by the Intergovernmental Panel on Climate Change (IPCC), established in 1988 by two United Nations organizations to make scientific assessments of relevant information on human-induced climate change. Accounts of the phenomena associated with it and their effects range from the Amsterdam Declaration of 2001, 'Challenges of a Changing Earth: Global Change Open Science Conference', to recent reports from the United Nations Environment Programme (UNEP) on the role played by climate change in the conflict in Darfur. A massive European Union project called PRUDENCE has produced specific projections of future temperature rises and drawn attention to their effects on the Paris region of a jump by the end of this century to temperatures of 30°C for 50 days per year compared to the current 6–9 days (Henson 2006: 53).

Such detection and attribution studies have the most lasting impact on public consciousness when set within the history of changes in the earth's atmosphere and climate, where present variations from norms set over millennia are graphically represented. The process of historicizing the earth that Martin Rudwick calls the 'reconstruction of geohistory' has been (and remains) a collaborative and broad-based enterprise. He traces its beginnings from the eighteenth century, when 'natural history', dealing with the description and classification of natural phenomena, natural objects and their diversity, was complemented by 'natural philosophy', dealing with the causal and mathematical relations between them (Rudwick 2005: 49–55).

A similar search for the most comprehensive possible knowledge of the earth's ecology informs the scientific body of work and ideas generated today by Gaia theory. Its approach is linked directly to the reconstruction of geohistory – the Greek name 'Gaia' given to the goddess Earth by Hesiod is echoed in the suffix 'geo'. Lovelock's theory has also helped move scientists and the general public toward a new understanding of earth history and, crucially now, *of our place within it*. This understanding is presupposed in the IPCC and other reports and in the routine professional and general use of the terms 'organism', 'ecology' and 'environment'. And, subliminally at least, in their now commonplace use within popular culture.

Such studies indicate that what is now happening to the climate is seen to affect us all, no matter where its immediate effect is detected – whether it be desertification in Africa, floods in Europe, migration of species northward or the melting of the Greenland ice cap. An implicitly and, now more often explicitly, inclusive world view is presented to us and reinforced through a globalized media, generating a globalized discourse that is used to express our common participation in a shared, ecologically understood life-world.

The scientific context

This discourse signals a transition from a postmodern perspective (in which the presumption of a 'grand narrative' of human history, or 'metalanguage', was discarded) to a scientifically grounded and shared story about the planet and its responses over time to our presence and actions. Global climate change (as the name infers) is recognized and chronicled as a universal story: told and retold in scientific language and then translated into common everyday speech and concerns (Myerson 2001: 18–37).

However, the danger with an exclusive use of this scientific narrative is, as we shall see, that it has presupposed our entitlement to handle global resources as though they are nothing but an inexhaustible supply of material for gratifying our desires – or indeed our curiosity. It has also presupposed that we have the competence to do so without damaging our future. Both these presuppositions underpin, to a large extent, the modern understanding of 'progress'. We are acclimatized to hearing and accepting the overall importance of 'growing' economic structures and institutions; of satisfaction at an expected rise in GDP and GNP and gloom at a fall; of welcoming expansions in travel and communications systems that keep us (potentially) in touch with a global clientele. Any threat to this growth, and the anxiety it arouses, is countered by government regulations, defence systems, border controls and tariffs that rely on surveillance techniques and, ultimately, on weapons whose lethal impact is sufficient to wipe out whole communities.

All the major elements in this world view rely on considering our future in purely human terms rather than as one common to all life on earth. Therefore, any real or perceived 'threat' to growth, to this ideology of continuous progress, is seen as a threat to both individual and collective self-interest. Against this background, an understandable, if unrealistic, reaction to the events of climate change and a growing understanding of their inevitable impact on lifestyles is denial. In a culture dominated by 'markets' and alienated from the natural environment, any notion of setting limits to their growth is seen as hostile to our progress as a species rather than as a proper reaction to their effects.

This results, in part at least, from the fact that over the past two centuries, without actually standing where Archimedes wished to stand – for we are still bound to the earth through the human condition – scientists appear to have found a way of acting upon the earth and within terrestrial boundaries as though from an Archimedean point outside it. So at the risk of endangering natural life processes, we have exposed the earth to forces alien to nature's household (Arendt 1958: 262) and have come, to some extent, to see ourselves as aliens there.

Both Arendt and her contemporary, Rachel Carson, were keenly aware of the 'fallout' from the use of atomic bombs in 1945. Carson linked this, and the release of the forces behind it, to the 'fallout' from pesticides and their enduring physical effects on the environment. Within the moment of time represented by the 20th century, she said, one species – man – has acquired significant power to alter the nature of his world:

> The most alarming of all man's assaults upon the environment is the contamination of air, earth, rivers and sea with dangerous and even lethal materials. This pollution is for the most part irrecoverable ... In this now universal contamination of the environment, chemicals are the sinister and little-recognized partners of radiation in changing the very nature of the world – the very nature of its life.
>
> (Carson 1962: 23)

This stands as testament to the potency inherent in events and to the truth of Arendt's comment on the character of that potency. Both despair and triumph, she said, are inherent in such scientific events as the release of energy processes formerly beyond human power or the production of elements not occurring naturally in the environment (Arendt 1958: 262).

The Christian context

Nevertheless, in spite of these and other concerns about the hegemony of science within Western culture it has, to a large extent, retained our trust in its ability to deal with those aspects of climate change for which it now openly accepts responsibility – such as the development and growth of fossil fuel usage, soil contamination, species loss and desertification. This public accountability (however flawed) is one reason why it enjoys the status of grand narrative formerly enjoyed by that of Christendom. This latter narrative, for reasons to be discussed at some length – such as its anthropocentric character, its legitimation of violence against the natural world and its emphasis on an otherworldly life that implicitly denies the value of this one – is generally viewed as no longer capable of meeting the multidimensional challenges posed by climate change. At this moment of transition and crisis within Western consciousness, religious language appears to have lost meaning or, even worse, to have inherited meanings that have grown perverse in the wake of a long list of modern atrocities that include world wars, genocides and nuclear armament (Robbins 2007: 3).

Yet because of its diversity of traditions, stemming from deep roots within the religious and secular histories of Eurasian and African countries and its transmission within and beyond Western culture generally, Christianity cannot simply be dismissed as incapable of revision without losing rich resources to help us face the challenge of climate change. Instead it can and must be re-envisioned in the light of present understandings of its history and of earth's history. A defining feature of that re-envisioning is a distinction between Christendom and a Christianity potentially based on a theology of nonviolence. For, to anticipate a little, the former refers to almost two millennia of the imperial power, authority and triumph of the Church. The latter, as we shall see, re-envisions the (almost) forgotten witness of Jesus to the nonviolent nature of the kingdom of God and his rejection of the kingdoms of this world and their glory (Matthew 4: 8–10).

The term 'Christendom' refers specifically to the imperial Christian religion that began when Constantine's Edict of Toleration in 312 CE ended nearly three centuries of sporadic, though at times quite severe, state-sanctioned persecution of Christians. By the end of the fourth century Christianity had become the dominant official religion of the Roman Empire and, with the exception of a few desert monks and later monastic settlements, the Christian Church became fully allied with the power of the state. This established a pattern of relations between Church, state and society that, for the next fifteen hundred years, ensured that most Christians learnt and practised their faith in the context of 'Christendom' (Robbins 2007: 4–10).

This has left us with a Christian theology that, in Catherine Keller's phrase, 'suffers from an imperial condition'. This is not, as she points out (and I associate myself with her statement), a denunciation lobbed from some indignant outsider, nor is it another hand-wringing confession or liberal hairshirt to be worn by theologians. It is first of all descriptive, indicating an organizational illness of history, deadly but not necessarily terminal.

> Empire is a recurrent condition, an extraordinarily adaptive one that grows rapidly in each new manifestation, voraciously consuming the space it occupies, mixing together otherwise separated cultures and ramming populations into relations extrinsic to their indigenous integrities.
>
> (Keller 2005: 113–14)

As I quote her, I can feel the nascent violence of the language she rightly uses in order to describe the event and effects of empire – it grows *rapidly*, consumes *voraciously*, *rams* cultures together. She concludes that the force of this imperialism has, from the start, made the work of theology global. Therefore, we need to consider the postcolonial potential of theology and the theological potential of postcolonial theory in our *constructive* projects that situate spiritual discourse within an irreducible sense of context. Then 'Christian globalism also and from the start translates into a *counter-imperial ecology of love*' (Keller 2005: 116).

A constructive theological approach

In this constructive frame of mind, the task of re-envisioning imperial Christian theology will necessarily concentrate on two focal points linked to its deep-rootedness in empire. The first is both an event and an idea within that culture: the creation and presumption of a single civilization under a single government. This connects with the contemporary globalization of culture and, in a time of climate change, to international government bodies (such as IPCC and UNEP) that enable worldwide research to be carried out aimed at the good of the whole. That this is being done in spite of a diversity of views and interests echoes the situation in Christendom where the Church was forced to admit into membership those who were not 'true' Christians and to condone contemporary values and customs that were unchristian.

The second constructive focus is on an important, inherently fricative element in imperial Christianity. This was (and is) the notion that there exists a small section of humanity specially chosen and loved by God. 'Choose' and 'elect' are translations of the same verb in the original Hebrew and Greek of the Bible. For the Jews, Israel was chosen to be God's people. In the Pauline writings the Church is called and chosen by God as the new Israel. But which Church? Claims by the present Pope that he leads the only true Church are characteristic of the tension between the faith of Christianity in the all-embracing love of God and the belief within Christendom that God has chosen an elect few.

Since the Protestant Reformation of the sixteenth century this has led to a steady dismantling of Christendom itself. It came to a head in the first half of the twentieth century when supposedly civilized, and nominally Christian nations turned against one another in total warfare. The shadow cast by this moral failure has greatly determined not only Christian theology but the very practice of contemporary religion.

Nevertheless, from beneath the shadow of this event there has emerged a prophetic voice whose words and observations are epitomized by Dietrich Bonhoeffer and his link to an

> emergent religious and cultural sensibility that was now forced to pick up the broken pieces and to imagine, if not craft, an alternative future – a future in the wake of the death of God and after the collapse of Christendom.
>
> (Robbins 2007: 8)

This is the religious sensibility that now has to craft that alternative future not only in the wake of the 'death' of a God of imperial power, but also beneath the shadow cast by the moral failure of our species to take responsibility for the effect of our actions on the future of all known life on the planet. One way to do this, I believe, is to build on a contemporary re-envisioning of what it meant, and means, to be a disciple of Jesus – rather than an imperial Christian or a self-styled member of the elect.

The move from Christian narrative to geohistory

This re-envisioning is being facilitated by the fact that Christians today are being forced to imagine an alternative future, for themselves and for the global environment, within a moral landscape whose cultural references are derived from a different grand narrative and world view than that of Christendom. Christian history is now recounted, studied and shared in a milieu dominated by scientific, economic and political discourses that have, for the most part, developed outside the categories of Christian thought. This common language is used in the Amsterdam Declaration, in the Stern Report, the IPCC Reports and notably in the 2002 UNEP Report on the 'Cultural and Spiritual Values of Biodiversity'.

The last mentioned highlights the need for a different moral landscape from one that would simply enshrine what scientific studies have discovered about biodiversity while neglecting to reveal its wonders. It questions a scientific method that still sees Nature as a collection of objects for human use, benefit and exploitation; that uses the banner of scientific 'objectivity' to mask the moral and ethical issues that emerge from a functionalist, anthropocentric philosophy; and that seldom embraces the values of local knowledge and traditions (Posey 1999: 5).

The UNEP Report does not hesitate to warn us about the future consequences of our actions not only for our children but for the planet itself. Nor does it hesitate to provide moral imperatives and definite religious contexts for reassessing the impact of those actions. Within this framework, the status of different theological narratives is at least provisionally accepted; but their orientation has shifted. For they are increasingly read, heard, acted upon and perceived, by Christian and non-Christian alike, against the backdrop of the 'grand' narrative of geohistory and a global concern with our role in it. The notion of an 'elect' within humanity, specially loved by God, is unequivocally called into question. For this newly emerging world view refuses to separate our well-being, let alone that of any particular section of us, from that of the whole community of life on earth.

This calls for some participatory thinking from those of us who are theologians. We must venture beyond the usual theological norms by which our individual futures are presumed to be determined by God and our conduct justified only before God. That means we cannot leave unaltered or skate over what is presently recognized and proclaimed as Christian theology. We must confront the role that Christian violence-of-God traditions have played in causing the problems raised by climate change and in justifying our part in them. Awareness of and attention to this state of affairs is a prerequisite for remedying it.

This awareness – and consequent commitment to change – is highly inconvenient for theologians. For one thing, being honest about it may simply reinforce the prevalent perception of religion as a violent force in society. But that perception must be faced and an alternative one offered by consciously choosing and concentrating on what is also already within the tradition: religious texts and teaching that do not sanction violence. Both honesty about Christian violence, and the decision to consciously focus on texts and teaching within the tradition that do not sanction it, will be employed here not simply as a critical exercise – although that is necessarily part of it. That honesty and decisiveness is part of a ground-clearing exercise from which to build nonviolent relationships between us and the whole community of life on earth. The climate this creates sustains a different kind of theological environment from that which has evolved around a God of power, sovereignty and punitive violence.

For me, as will become clear, that has meant developing a way of thinking about God within the mystery of giving. This in turn changes our Christian self-perception to one that sees us as ineluctably situated within a comprehensive narrative of gift events and exchanges throughout earth's history: a much larger scenario than the time-specific, separate and sectarian theological narratives of a personal salvation history.

Such a participatory, rather than an anthropocentric approach has the potential to transform our largely utilitarian theological world view into one shaped by concern for the welfare of all beings and attentiveness to their suffering. A utilitarian religious perspective habitually uses violent images of God from sacred texts to sanction oppressive relationships with the earth's inhabitants and exploitative attitudes toward its life resource systems. Acknowledging that such violence has been, and is, perpetrated in the name of a transcendent, omnipotent and disembodied deity alerts us to the need for a nonviolent and embodied vision of our relationships: whether with God or with the other members of life's community.

Moving towards this means examining how sacred images and stories have been, and still are, used in our literate culture for violent purposes. One reason for this usage is that, implicitly at least, they appear to issue from an unmediated, transcendent vision of the earth and its inhabitants that is situated in neither place or time, nor in a particular human community. Such a vision, bound (or rather, unbound) within 'absolute' dimensions of sacred or salvation history, superimposes an 'unearthly' image of land, people, animals and resources onto and over our actual relationships within real landscapes and communities. And because this salvific vision transcends our individual lives it appears to relieve us of responsibility for the effects of our actions on others. Max Weber called this religious attitude one of 'acosmic indifference to obvious practical and rational considerations' (Weber 1965: 195).

More grounded religious cultures, however, rely on situated knowledge: on recursive layers of stories and metaphors that tie people and land together in inter-connected networks validated through the experience of ancestral generations. Biblical studies reveal this multi-layered process at work within the written texts. However, they are routinely read by Christians as the voice of a God whose con-stitutive relations with the world are those of an unaccountable, unlimited, atemporal and therefore unchanging power: totally other than any power experi-enced or wielded by any person. In classic Christian terms, formulated by Anselm, 'all necessity and impossibility is under his [*sic*] control' (Blond 1998: 208).

This biblical 'overview' is an illusion based on the concept of an infinite vision attributed to an omniscient, non-locatable God and revealed only to an elect. It appears to offer unsituated, unlocatable knowledge about the historical integra-tion of humans with each other. Furthermore, these claims to privileged knowledge are extended to the relationships between them, their lands and the multitude of nonhumanss conventionally called Nature. And because these claims appear to issue from a realm outside finitude, error, suffering and death, acting upon them allows those who invoke them to escape accountability for their effects (Haraway 1997: 131–8).

In a time of climate change, however, this biblical overview is challenged by a general acceptance that, for instance, our continued overuse of fossil fuels and other finite resources will inevitably lead to a further rise in global temperatures. And the latest IPCC Report highlights, for those of us whose own lives may not be directly at stake, that this rise will have dire consequences for the lives of the poorest, most marginalized groups of people, their environments and all life forms within them. Most tellingly, it also lays the responsibility for those conse-quences squarely at our own door.

2   The Seminal Event

God created
That is the novelty ...
Here the shell of the mystery breaks

(Rosenzweig 1985: 112)

In the previous chapter I contrasted two different ways of understanding God. The first envisions God as transcendent, unaccountable, omniscient, all-powerful and non-locatable in relation to the earth. The second perceives all understanding of God as emerging from earthly knowledge and firmly situated there. While the two may coalesce in different ways at different times, it is the second that will help us respond positively to climate change. It presupposes that God, the people, the land and all its inhabitants coexist in interconnected networks of life-giving relationships validated through the experience of ancestral generations.

The theological event

How does this earth-based understanding of God accord with the Genesis account of Caputo's 'seminal event of creation'? Franz Rosenzweig, who collaborated with Martin Buber on a translation of the Hebrew Bible, reflected throughout his life on the Genesis text and on what it disclosed of 'Creation, or The Ever-Enduring Base of Things'. His use of the term 'novelty' in the epigraph to this chapter indicates the totally new, unpredictable and, therefore, mysterious nature of the emergence of earth and skies. It has all the innovative, productive and life-giving character of Caputo's 'seminal' event.

Rosenzweig's lengthy reflections on the Genesis text and the creative process it describes – albeit in a slightly different sequence – are summarized by him as follows:

The beginning was: God created.
God created the earth and the skies.
That was the first thing.
The breath of God moved over the face of the waters:

Over the darkness covering the face of the deep.
That was the second thing.

Then came the third thing.
God spoke.

God's vitality, said Rosenzweig, transforms itself into a beginning, into an enduring basis for all things existing on earth and in the skies. (Rosenzweig 1985: 112–19). Catherine Keller echoes this thought when she says that the term 'creation' emphasizes the creative novelty, the mysterious event-character of what comes into being. We cannot simply exchange self-contained words such as 'universe', 'cosmos', or 'nature' for the word 'creation' because they omit the relationship to a creator embodied in the events they signify. 'Creation,' however, is a term that has been barnacled with stereotypes of a great supernatural surge of unlocatable father-power. And in relation to that 'Him', 'the darkness covering the face of the deep' has been identified in dominant Western theology as an evil, revolting chaos to be mastered or a nothing to be ignored.

Keller quotes a speech before the whole US Senate in 1900 by Senator Beveridge that has unholy echoes today: 'God has made us the master organizers of the world to establish system where chaos reigns. He [*sic*] has made us adept in government that we may administer government among savages and senile peoples.' From the vantage point of the white colonizer, the evil is always disorder rather than an unjust order; anarchy rather than control, darkness rather than pallor (Keller 2003: 5–7).

James Baldwin attested to the everyday effects of this:

> I learned in New Jersey that to be a Negro meant, precisely, that one was never looked at but was simply at the mercy of the reflexes the color of one's skin caused in other people.
>
> (Baldwin 1990: 93)

These stereotypes emerged from (mostly) patriarchal Christian communities and coloured their interpretations, descriptions and claims to understanding the creation event. They continue to do so, even though that beginning, that necessary seminal event or 'first thing' we call creation, occurred at a time when we were not there to record its progress. It happened when there was no time. It names a horizon forever cloaked in mist, one where our questions fade and no answers return to us. For it began without us (Drees 2002: 14).

At best, we and our ancestral generations reconstruct that 'time when there was no time' as the moment of a seminal gift event. From its primary givenness, its creative novelty, all other life-giving events have since flowed. From and within this enduring base, its elements – the womb-like deep (*tehom*), the 'formless void' (*tohu wa bohu*) and the spirit (*ruach*) – were eventually transformed into an order of coexistence from which would emerge the continuing process we call the evolution of species.

But, as God reminded Job, we were absent. It happened without us. All too often this is forgotten when we give our accounts of it – and of its creator:

> Were you there when I stopped the waters,
> As they issued gushing from the womb?
> When I wrapped the ocean in clouds
> And swaddled the sea in shadows?
> When I closed it in with barriers
> And set its boundaries, saying,
> Here you may come, but no farther;
> Here your proud waves break.
>
> (Job 38: 8–11)

This reconfigures creation as an image of 'natality', of a new beginning inherent in birth that makes itself felt in the world because it possesses the capacity of beginning something new (Arendt 1958: 9). And note how this imagery of the vitality and nascent potential inherent in the seminal event of creation is surrounded by silence in the original text in Genesis 1: 2. There is no word of command. It was only *after* this, as Rosenzweig points out, that God spoke. By the time of Job, however, that commanding voice was continually ascribed to a God with transcendent power over, rather than within, the creative event.

Keller uses the image of natality in the Job text in a moving and concise reconfiguration of Genesis 1: 2. The ocean bursts forth in amniotic liquidity, caught like a baby by the midwife and wrapped in the soft darkness. This does not diminish the mythic danger of the sea and the darkness of origins but it liberates them from the mood and meaning of evil. The sea's ferocity is portrayed as the agitation of an infant that requires boundaries for its own protection. She finds deeper echoes of these maternal impulses in the text where Jacob blesses his son Joseph:

> Blessings of the skies above,
> blessings of the deep that lies beneath,
> blessings of the breasts and of the womb (*rehem tehom*).
>
> (Genesis 49: 25)

But then whose womb is this that precedes all creatures? From the perspective of the whirlwind, the *ruach* that pulsed over the deep, how can we avoid the inference that the *rehem* is God's, from whose unfathomable *tehom* the waters issue? (Keller 2003: 130–1).

The philosophical event

Rosenzweig saw God's vitality transforming itself into both a beginning and an evident and enduring basis for all things in creation. In what can be read as an expansion of the bi-polarity of this thought, Gilles Deleuze comments:

In the one case, we move from the cosmic present to the not-yet actualized event; in the other, we go from the pure event to its most limited present actualization. Moreover, in the one case we link the event to its corporeal causes and to their physical unity; in the other, we link it to its incorporeal quasi-cause.

(Deleuze 1990: 164)

From this perspective, the pure seminal event of creation releases what Rosenzweig terms its divine vitality into present actualization. Similarly, Caputo sees its potency and Deleuze its internal causality actualized in two ways noted by Rosenzweig. It turns our eyes outward, to what has been and is being created. At the same time, the mystery inherent in this creating, the incorporeal quasi-cause, becomes manifest. Yet, by its very nature, this means that it exceeds all attempts to describe its present state definitively and resists any attempt to reduce it to the limits of human understanding.

Also, because this event moves in two directions at once, between past and future, it always eludes the present, causing future and past, more and less, too much and not enough to coincide in the simultaneity of what is manifest in creation (Deleuze 1990: 3). It is this, agrees Caputo, that makes an event an irreducible possibility; a potentiality that can assume any number of varieties of expression and instantiation (Caputo 2007: 51).

Philosopher Alain Badiou highlights another important characteristic of an event.

It is both *situated* – it is the event of this or that situation – and *supplementary;* thus absolutely detached from, or unrelated to, all the rules of the situation.

(Badiou 2001: 68)

Situating the event of creation within 'the earth and the skies' keeps our attention firmly fixed on what is accessible to sense perception. We know what we are referring to, can even point to it if we so wish. (This relationship between the situatedness of creation and our perception of events within it will be explored at some length in Chapter 5.) At the same time we are aware that creation is 'detached' from all earthly rules of beginnings and endings because of its irreducible possibility. We might then ask, says Badiou, what it is that does make the connection between the event and that for which, and in which, it *is* an event.

His answer takes us back once again to the *tohu wa bohu*, 'the formless womb-like void' of Genesis. The connection, he says, is the situated void (*vide*) of the earlier situation. He gives as an example the emergence of the classical style of music from the baroque, a situation in which there was a void, an absence in which a new, supplementary way of developing musical writing could appear. Yet, because of its situation in relation to the baroque, that void could itself be named as a new form of *musical* writing. What any event reveals – and the truth of this will emerge in the following chapters – is that something in 'the situated void' had its own identity that was not, indeed could not be taken account of in the earlier situation (Badiou 2001: 69, 134). One event may be seen as the birthing place, the womb from which others emerge.

This brings me to an important distinction made by Caputo between the name and the event, or the thing and the event. Or indeed, between the name, the event and its effects. As we have seen, naming the seminal event 'creation' in no way sums up its meaning, its remit or its future course. Nevertheless the vitality, the potency in the event of creation and, in particular, in its drawing our eyes outward, continues to lure us into describing its past and imagining its future. So earlier descriptions of creation offered by the biblical authors have now been augmented, and for many, replaced and surpassed by scientific accounts of its vital processes.

The scientific event

Contrary to the biblical account, scientists situate, relocate and by doing so, rename the seminal event of earth's creation as the origin of the universe. They are concerned with the universal origins of earth and sky. And note how effectively this has widened the scope of our ignorance! We were certainly not there at that point outside our time when there was no time.

Nevertheless, using data gathered with the help of increasingly sophisticated technology, scientists are able to piece together a contemporary narrative of earth's creation that resulted from the seminal event that created the universe. James Lovelock names this emergent event and its effects 'the birth of Gaia'. I shall summarize its narrative arc here to show how correspondences may be discerned between it and the Genesis creation story.

Our earth was born about 4.6 billion years ago after a supernova exploded close to clouds of hydrogen and helium. This thermonuclear explosion caused the clouds to condense under their own weight, caused elements to be synthesized and the sun and planets to be formed. Fierce winds from the coalescing sun flung out the emerging planets, whose collisions with other debris caused fragmentation and moons to appear. Our molten planet was born and began to cool.

This corresponds to what Genesis calls 'earth'.

For a billion years, this planet was a melting pot of chemical reactions, evolving gases that became its lasting atmosphere. There were many active volcanoes that surrounded it with fiery turbulence, the molten basalt thrown up reacting with the gases to form the first rocks. Carbon dioxide, water vapour, nitrogen and carbon monoxide blanketed the planet's surface.

This corresponds to what Genesis calls 'skies'.

As the planet cooled, water condensed and collected in the pits and hollows of the planet's surface. At a particular period, the chemistry, temperature and climate of the earth became favourable for life. Somewhere, somehow, compounds of carbon and of other elements formed that were stable enough to react further and form ensembles of compounds and structures. During this period, the first micro-organisms emerged that, deep in those waters where volcanic fissures erupted, captured and retained the hydrogen produced during the reaction of water with basalt rocks on the sea floor. Some micro-organisms learnt to photo-synthesize and this released oxygen that, reacting within escaping hydrogen, produced more water. Thus our planet, unlike Mars and Venus that in early times had water in abundance, retained its oceans and did not become irreversibly dead (Lovelock 1991: 74–80).

This corresponds to what Genesis refers to as 'the breath (ruach) of God hovering over the waters'.

In these expanded accounts of what we now think happened in that seminal event, the shell of the mystery opens fractionally wider to allow the smallest glimpse into what will always remain essentially hidden. Both religious and con-temporary scientific descriptions tell us that the 'first things' to emerge from it were earth, skies, water, life, breath. Without their interactions and interdepen-dence the earth could not have brought forth the grass, the seeds, the trees and their fruits. Without their interactions the waters could not have brought forth a swarming of living souls. Without their interdependent co-arising earth could not have brought forth creepers on the ground, beasts feeding on plants and creatures flying in the air. Without their continuing interactions and interdepen-dence, we would not be here today.

Remembering these 'first things' antecedent to our own emergence properly awakens a sense of awe in us. In response to the account of them given by God to Job, he can only stammer: 'I have uttered what I did not understand, things too wonderful for me, that I did not know' (Job 38: 1 to 42: 3). This inade-quacy has been routinely felt and expressed by scientists too, as they study the processes by which living organisms continue to emerge and adapt themselves to the conditions of their global environment and, in turn, transform it by the way they live.

In 1943 Nobel laureate Erwin Schrödinger posed the question of how physics and chemistry can account for the events in space and time that take place within the spatial boundary of a living organism. Thanks to the work of biologists, he said, enough is known about the actual material structure of organisms and about their functioning to state why, and precisely why, present-day physics and chemistry could not possibly account for what happens in space and time within a living organism. The reason, he said, is that the laws of physics and chemistry are statistical throughout. And it is in relation to the sta-tistical point of view that the structure of the vital parts of living organisms differs so entirely from that of any piece of matter that physicists and chemists have ever handled physically in laboratories or mentally at their desks (Schrödinger 2000: 4).

Updating our knowledge of these processes – and of our role in them

Today, all three branches of science certainly offer a fuller account of what happens *between* as well as within living organisms and now, within and between them and micro-organisms. It is now the case that molecular biologists, for instance, have advanced our knowledge of living organisms by studying their symbiosis: the interactive living together of very different kinds of organism that, in certain cases, results in symbiogenesis: the appearance of new bodies, new organisms, new species (Margulis 1998: 33). Nevertheless, Schrödinger alerts us to the dangers inherent in an Archimedean point of view where the structures of living organisms are considered as if from a point external to their existence.

However accounted for or described, we know that these bodies, like our own, receive and utilize energy from the sun and from each other, as part of a constantly recurring, self-regulating, self-purifying, complex adaptive process. The interactions between them, together with those between them and the physical and chemical environment of the planet, have modified conditions within it, creating more ecological niches for the emergence and continued existence of many types of living being. The complex dynamics of this process have created life support systems that intimately affect all living beings and are themselves dependent on interactions between receivers and givers. Green plants, for example, fix nitrogen and synthesize tissue containing stored chemical energy, an essential food source for a vast number of animals – including ourselves – who cannot do it for themselves. Directly or indirectly, the energy driving these ecological cycles flows from the sun, and the diversity of life within them assures their resilience.

Our present knowledge and experience of the processes set in train by the event of creation is spelled out in some detail in *The Encyclopaedia of Life Support Systems (EOLSS)*, a web-based sustainable development archive, supported by UNESCO, with a current knowledge-base of global life support systems' relevance (www.eolss.net). In a series of studies of different aspects of these systems it makes many important points about the part we play in sustaining – or not sustaining – them. I want to highlight two of those points.

The first is that the fundamental attribute of life support systems is (as the name infers) that they work together to provide all that is needed for the continuance of life. These needs, the authors state, go far beyond biological requirements, encompassing *natural environmental systems as well as ancillary social systems required to foster social harmony, safety, nutrition, economic standards, development of new technology, etc.* [Italics mine]

That concluding 'etc.' highlights a second important point to emerge from the *EOLSS* studies. They assume that the multitude of life support systems, too numerous to enumerate, categorize or investigate because of their essential variety and interactive nature, are located within our earth-space. They flow from, and depend on, the earth-body that is the ground of shared existence. Therefore, they exceed all our attempts at adequate description, definition or expression, whether theological, scientific or symbolic. It is this that makes the process of giving and receiving life, considered as a whole, a mystery to us. Our perception of it as a whole is always limited by our being part of it.

Wittgenstein expressed one aspect of this truth when he said that it is not *how* things are in the world that is mystical, but *that* it exists. And that viewing the world *sub specie aeternitatis* is to view it as a whole – a limited whole. This limitation is linked, he said, to the fact that for each of us the limits of *language* (of the language that I alone understand) mean the limits of my world (Wittgenstein 1961: 73, 57).

Nevertheless, as we experience and become aware of our dependence on, and interactions with, these all-encompassing life support systems, we need to know more about how they do what they do. At a time of climate change, we categorize and investigate them, not simply in order to master or control them but to try and grasp our role in them by classifying, interpreting and learning to understand their products and effects. Their global reach is, implicitly at least, perceived and described – however inadequately – in a variety of ways. Their products can now be classified as emergent phenomena: that is, as life forms, events or experiences that resist explanation in traditional cause-and-effect sequential language.

Reacting to the seminal event

Interweaving reflections on the language of the Genesis text with that of science enables us, in some measure, to grasp the connections between the seminal event of creation and the nature of multiple and scattered events, distant and close ones, that affect us and our own situation on the planet in multifaceted ways. But tying in scientific explanations with a prior religious understanding that issues, however minimally, from a cultural familiarity with the biblical narrative, does not mean infusing religious meanings into scientific discourse or collapsing one discourse into another.

It does mean, however, recontextualizing biblical accounts. That poses a particular challenge to theologians. How do we change our view of the relationships between earth, ourselves and God as a result of a change in data about the seminal event? How have scientific findings changed our image of ourselves and of our place within those relationships?

Those findings, and the fact that they too change, can evoke confusion and resistance in us as they challenge our confidence about our place in the world and its practical implications. They can also, however, evoke a different kind of religious response in us: a sense of connectedness with all living beings that, to some extent, counters our feeling of being resident aliens on the planet, displaced from our true home in heaven. Geneticist David Suzuki quotes Albert Einstein:

> A human being is part of the whole, called by us the universe. A part limited in time and space. He experiences himself, his thoughts and feelings, as something separate from the rest, a kind of optical delusion of his consciousness.

This delusion is a kind of prison for us, restricting us to our personal desires and to affection for a few persons nearest to us. Our task must be to free ourselves from this prison by widening our circle of compassion to include all living creatures.

(Suzuki 1997: 26)

The problems attached to a separatist view of ourselves, from seeing ourselves as separated from all other life forms by our intelligence or presuming our separate creation by God, will emerge in a variety of forms. The question here is how our present engagement with scientific data affects the meanings we attach to the phrase 'seminal event'.

Returning briefly to the semantics of 'event', it is not confined to what happened or happens – which is what the word 'event' suggests in English – but signifies something going on in what happens; something that is being expressed or realized or given shape in what happens. The seminal event of creation is not something present but something seeking to make itself felt and known in what is present. Accordingly, Caputo distinguishes between the *name* (for example, 'creation', 'symbiogenesis' or 'climate' as a name for a universal phenomenon) and the *event* (natality, symbiosis or climate change) that 'is astir and transpires in a name'. The name is, he says, a kind of provisional formulation of an event, of a relatively stable if evolving structure; while the event is ever restless, on the move, seeking new forms to assume, seeking to get expressed in still unexpressed ways. Events get realized in things, take on actuality and presence there, but always in a way that is provisional and revisable.

In terms of their temporality, events, never being present, solicit us from afar, draw us on, draw us out into the future, calling us hither ... Or else they call us back, recall us to all that has flowed by into the past ... Events call and recall.

(Caputo 2007: 47–8)

Whether naming an event 'natality', 'climate change' or 'symbiosis', looking back recalls all that has flowed into the present from births, climate or symbiogenesis within the earth system as a whole. The word 'Gaia' recalls a time in prehistoric Greece when this was the name for the earth as goddess. Now it is astir in a theory that takes us back further still, to the events over three billion years ago in which earth first displayed the capacity to stay close to the temperature and chemical composition that created a climate able to bring forth and nurture life.

The name Gaia also brings us up to the present when the event of this self-regulation is seen to maintain earth's climate in a state of dynamic equilibrium that enables us to describe her as a living organism (Lovelock 1991: 29, 141; Lovelock 2007: 47). Looking forward, it draws us out into a future where the global climate that emerged from these Gaian feedback systems and planetary relational processes is so compromised by human-led events that it may not continue to provide the life support that makes our lives, and indeed all other lives, possible – at least in the way we experience life today.

The third thing: God spoke

Theologically, the enduring event we call creation harbours the name of God. What is astir and transpires in that name for us today?

In the Genesis text the focus of attention is on the God who gives commands, commands to all living souls to be fruitful, to fill the waters of the sea and to increase upon the earth. When this command is given to the human male and female, it is accompanied by the command to 'fill the earth and subdue it'.

According to the statistics on human population figures they are now at an all-time high: approaching seven billion and rising. It is salutary and sobering to reflect on how that absolute command has resounded and continues to resound in some Jewish and Christian women's ears. And how its resonances persist in concerns centred on human fertility in the developed world at a time when our relentless subduing of the earth and destruction of its life support systems can sustain far fewer numbers of us.

But this God of Genesis allows no appeal against his commands. He wields absolute power over the life and death of earthly bodies, whether vegetable, animal or human; while having no body himself. This God is only a voice, an unearthly and therefore disembodied voice. The granting to bodies of life, of fertility, of fruitfulness, of knowledge, as well as the imposition of pain, death and exile is recorded as an act of disembodied power. The voice of command is literally that, and nothing more. There can be no appeal to compassion, to bodily experience or to identifiable emotion against the edicts of that voice. Its pattern of command-giving is that of male imperial power, enthroned above and beyond human emotion or weakness. And the history of Christendom has reinforced this perception of the power attached to the name of God.

Therefore, no matter how we try, we cannot make the names and images of God presented to us in the Bible internally consistent or indeed universally malevolent or benevolent. Nor may we use images of God acting as an imperial despot to justify violence, patriarchy or destruction of earth's life support systems. We are obliged to see the Bible and the events recorded in it for what they are: a mixed collection of texts from different times and places that give us, if we so choose – and choose we must – images that foster violent or nonviolent relationships between us. And when the image highlighted is one that may legitimate violence, there is an added obligation to be explicit in our rejection of it and our choice of one that encourages nonviolent cooperation. And we need too to be explicit about our reasons for making that choice.

Such a choice covers a wide field but my focus in this book is on texts centred on events in the life of Jesus. What God do they harbour? Within the Hebrew Bible too, there are many options open to us within the same narrative tradition. Within the Pentateuch, anthropologist Mary Douglas offers us a powerful, indeed startlingly constructive reading of the book of Leviticus, where the religious rules and rule-keeping governing the community offer us what today is a counter-cultural vision: that of an ecologically sustainable society.

Within that society divine injunctions to be compassionate, for example, would not be necessary because kindness is predicated in the rules of behaviour as well as exemplified in the narrative. The book can be read in parallel with Psalm 145: 8–9. The God of Israel has compassion for all he has made. His love for his animal creation lies behind laws that legislate for justice between persons, between God and his people and between people and animals.

Being moral would mean being holy as God is holy: being in alignment with the universe, working with the laws of creation that manifest the mind of God:

> You have made the moon to mark the seasons;
> The sun knows its time for setting.
>
> (Psalm 104: 14–29)

Her argument for this reading and how its rules worked rests on her account of the society as one imbued with analogical-rational rather than rational-instrumental thinking. The latter creates contexts in which 'human nature' or 'human rights' can be ordered or legislated for within a linear, hierarchical model. Analogical thinking, however, is multi-threaded rather than sequential. It places an item or event within a scheme organized in terms of analogical relations among the items selected and then reflects on them and acts in terms suggested by the relations. The meaning of one item or element:

> Would now always have to be sought in whole systems of meanings ... No more is it useful to consider what separate items such as bread or blood symbolize. It is not just a change in our methods of work, but a change in our understanding of human thinking ... What is important is not the one-to-one relations of group to species but the overall confrontation of human society and nature.
>
> (Douglas 2000: 22–44)

This contemporary reading gives some idea of a society that honours the order it finds implicit in the seminal event of creation and, by keeping its laws, shares in and cherishes the vitality of that event. For its biblical author, 'the living body is its paradigm' (Douglas 2000: 35, 37).

In the following three chapters I shall look at some historic events much closer to us in time, and at their continuing effects on our present situation in a time of climate change.

3 The First Historic Event

The problems regarding colonization can be stated above all in terms of force. Colonization nearly always begins by the exercise of force in its purest form, that is, by conquest ... Christ never said that warships should accompany, even at a distance, those who bring the good news. Their presence changes the nature of the message

(Weil 2003: 66)

The seminal event of creation both precedes and continually enlivens human history. Within that history Hannah Arendt singles out three particular events that are antecedent to every life today yet present in the 'givenness' of each of our lives now and in the future. So each in its own way can be seen as contributing to the climate change crisis. They are: the discovery of America; the Reformation and the invention of the telescope.

These stand, she says, at the threshold of the modern age and determine its character. In the eyes of their contemporaries the most spectacular of these events must have been the discoveries of unheard-of continents and undreamed-of oceans. The most disturbing might have been the Reformation's irremediable splitting of Western Christianity with its inherent challenge to orthodoxy and its immediate threat to the tranquillity of their lives and souls. Certainly the least noticed was the addition of a new implement, the telescope, to an already large arsenal of tools. Useless except to look at the stars, it was in fact the first purely scientific instrument ever devised in the West (Arendt 1958: 248).

In the following chapters I shall look at different aspects of these three events and assess some of their present effects, with particular emphasis on their power to transcend the limits of a particular era and, therefore, the impossibility of reducing what happened in them to their effects at any one time. As is the case with past events, from our present perspective despair and hope appear inherent in each of them.

The beginnings of European colonization

Bearing in mind the previous colonizing of Africa and the Middle East by the Greeks and Romans, chronologically the first of these three events was the discovery of America and the ensuing exploration of the whole earth, with the mapping of its lands and the charting of its waters. The shrinkage of the globe that followed has made each of us as much an inhabitant of the earth as of our country. For we now live in an earth-wide continuous whole where even the notion of distance has yielded before the onslaught of speed of access. We know we can reach countries far distant from our own in a matter of hours rather than days or months.

This physical shrinkage has been accompanied by another, infinitely greater and more effective, brought about through the extended technical surveying capacity of the human mind. Numbers, symbols and models are used to condense and scale earthly physical distance down to the size of the human mind's natural sense and understanding. Of course when Arendt wrote, the invention and use of the Internet was yet to come, a tool reinforcing and epitomizing the shrinkage of the globe in the speed of our access to all its parts and to those who inhabit them.

She was, however, concerned with an aspect of this technical capacity to survey the earth that is also now of even greater importance. It functions, she said, only if a person disentangles himself from all involvement in and concern with what is close at hand and withdraws from everything near him. This echoes Schrödinger's remark that present-day physics and chemistry consider the structures of living organisms from a point of view external to their existence. The greater the distance from what is close at hand, the more we are able to survey and measure. Therefore, the most decisive physical shrinkage of the earth has followed from the invention of the aeroplane and from there to today's opportunities to travel at subsonic and supersonic speeds. All of which is enhanced by the speed and ease of Internet and satellite communication systems.

This has particularly affected our relationship with that part of the earth that is close at hand and immediately supports our existence. For we now have the technology to enable at least some of us to leave the surface of the earth and to view it, and ourselves, from a point external to our normal existence. This is an intensification of a process that, Arendt said, symbolizes the

> general phenomenon that any decrease in terrestrial distance can be won only at the price of putting a decisive distance between man and earth, of alienating man from his immediate surroundings.
>
> (Arendt 1958: 250–1)

While one might now wish to adopt a less gender-specific language than that used by Arendt, *pace* women astronauts it is correct historically. By virtue of those who initiated, contributed to and participated in them, the events and advances in science that she notes, as well as that of the Internet in our time.

were exclusively male-dominated enterprises and have remained so to a large extent. The history of Royal Societies, learned Academies and global Institutions records this as fact.

Theological colonization

Returning to her historic starting point, the discovery of America and the ensuing exploration of the whole of the earth's surface, we can see that not only did this potentially and eventually transform us into global citizens but also, through increasing emigration/immigration flows, into resident 'aliens' on whatever patch of earth we now choose (or not) to inhabit. More immediately, it signalled the beginning of European Christian colonialism on a global scale. That, and the events surrounding the colonizing process, had a profound and extensive influence on world history and lethal effects on whole peoples and on the biodiversity of their environments.

Theologically, in the name of the Christian God, it validated slavery and displacement of peoples, appropriation and exploitation of their lands and eventually massive depletion of natural resources. In the 1660s, a founder member of the Royal Society in London, John Evelyn, addressed the problem of the conservation of English timber by recommending the removal of most iron mills to New England, lest they 'ruin Old England'. Conservation at home was thus to be purchased at the expense of frontier expansion and forest clearances abroad (Merchant 1980: 236–9).

Such negative effects on colonized lands marked and still marks Western Christian expansion worldwide. With it went a belief that collective identity is forged by a monotheistic 'covenant' *against* the claims of others to a different identity. This is loudly and eloquently proclaimed in documents such as Christopher Columbus's diary on his first voyage west in 1492, in the Scottish Covenants of 1638 and 1643 and even, according to some interpreters, in the US Constitution. Any alternative vision of monotheism that offers glimpses of multiple identities or of a God of endless and inexhaustible giving is almost impossible to maintain in the face of a colonization where an identity negatively asserted is bound up with competing claims to possession of land (Schwartz 1997: 25f.).

It is most likely too, as Enrique Dussel points out – but that doesn't make the results any less depressing – that the real motives for the conquests can be traced to the economic and political expansionist projects of the trading powers within 'Christendom' (Dussel 1990: 61). Those powers have always been linked directly to economic policies and polities that have contributed to, and continue to contribute to, such factors in climate change as depletion and overuse of raw materials and energy resources available as mineral deposits, fish 'stocks' or forests. Western Europe thus became the primary beneficiary of a world economy, mainly by controlling interregional maritime trade and exploiting natural wealth on a world scale. This meant that cheap resources flowed into Western cities while the demand for European goods soared with colonial expansion. As a result, large segments of the European population have enjoyed a higher standard of living than they would have otherwise (Richards 2003: 17f.).

This has contributed to a problem already mentioned that arises from connecting such projects to a belief in, and an emphasis on, 'progress.' In the various post-colonial forms of domination whose effects continue to plague the world, Western powers continue to insist upon lauding their progress in such matters as industrial technologies, computer-aided design, speed of travel and communications systems. Through such financial instruments as the World Bank and the IMF these are imposed on countries seeking monetary aid. The downside to all this is the continuing exercise of global power under the guise of progress, whether economic, military, cultural or, as in the case of the Iraq war, on the grounds of imposing progressive, that is, democratic government. In all this expansion of power the use, influence and effects of massive military expenditure and global communication systems cannot be underestimated.

Simone Weil observed that from the point of view of the colonized, Western culture, dressed up in its own prestige, its model of progress and of victory, always manages to attract a proportion of the youth of the colonized countries. Technology may at first offend traditional ways, but it subsequently seduces by its power. The colonized populations, at least in part, want nothing more than to assimilate that culture and that technology. If this desire is not immediately manifest, it comes about almost infallibly with time (Weil 2003: 67).

The truth of her observation is evidenced today not only in the French colonies but in the Internet cafes and trade in mobile phones throughout shanty towns and slums in Africa, Asia and South America. This 'urbanization by stealth' is accompanied by a flight from the land and loss of indigenous farming methods. By implication and in fact, Western economic progress has been at the expense of a majority of people, both outside and within Europe.

Viewed from a theological perspective sensitized by the inconvenient truths of climate change, it must also be acknowledged that this economic progress was validated by a Christian militarist categorizing of the indigenous inhabitants of the lands they colonized as *enemies* of God – Muslims, heretics, pagans and idolaters (Dussel 1990: 39). The genocide and abuse of their bodies and exploitation of their lands would not be possible, either then or now, without an implicitly militarist theology: one in which physical violence is linked to theologically violent language. Today our ears ring with defamatory, aggressive religious rhetoric about 'the axis of evil' and those who belong within it; both assertions countered in similar terms by those so categorized.

Effects of violence-of-God traditions

An important consequence of validating this violence theologically is a refusal to accept accountability for the suffering it causes. This is most acute in respect to the suffering caused to earth's nonhuman inhabitants. While there is an ongoing unofficial effort to maintain an Iraq 'Body Count', it is, as far as I know, limited to human bodies. The suffering and death inflicted on nonhuman ones is literally discounted.

Their suffering can appear to be, and indeed is, sometimes explicitly condoned as the divine will. One sacred text that has God saving a chosen group of people from land flooded by a catastrophic weather event – that would now be attributed to climate change – exemplifies religious sanctioning of violence against its nonhuman inhabitants. Its conclusion now reads as a self-fulfilling prophecy:

> And God blessed Noah and his sons and said to them: 'Be fruitful and multiply and fill the earth. The fear of you and the dread of you shall be upon every beast of the earth, and upon every bird of the air, upon everything that creeps upon the ground and all the fish of the sea; into your hand they are delivered.'
>
> (Genesis 9: 1–2)

What image of God does this depict? One who has supposedly delivered all known life forms into our hands while condoning, indeed affirming, their well-earned fear and dread of us. This is a God who endorses, indeed imposes, that fear and dread on them and takes no account of their well-being, suffering or disappearance. So we cannot be held accountable for it. For such a God is made to appear as blessing our multiplying human populations and resource-expensive economic cycles of production, consumption and waste.

In a time of climate change, belief in such a God goes against everything needed to avert catastrophic weather events and their devastating consequences. For it supports what Max Weber called 'an acosmic indifference to obvious practical and rational considerations' (Weber 1965: 195). Those considerations are spelt out and underlined in every scientific survey and projection of the effects, for instance, of positive feedback fuelled by increasing carbon emissions. Never before have humans been made so conscious of the future effects of present actions.

Yet a transcendent vision of ourselves (or of some chosen section of us) as uniquely blessed by God, to the exclusion and the detriment of all other life forms, is understandably well-received, argued for and acted upon in traditional as well as fundamentalist religious communities. Proclaimed by some colonizers as their 'manifest destiny', that is, as the salvation by God of an elect few on the model of the election of Israel, it has helped to justify the near-extinction of indigenous peoples and of native species whose present parlous state is exacerbated by the fact that their existence has always depended on occupying precise ecological and climatic niches. By contrast, we, through our ability to adapt to diverse habitats and climate conditions by building arks, bridges and aeroplanes, continue to 'multiply and fill the earth' in unprecedented numbers.

Jack Nelson-Pallmeyer makes a neat connection here when he says that the 'inconvenient truth' of the troubling realities of climate change ought to alert theologians to the 'inconvenient truth' that certain readings of sacred texts – such as the Genesis text above – and traditional teachings based on them, have both provided and sanctioned violent images of God that have in turn sanctioned religious violence of all kinds (Nelson-Pallmeyer 2007: 9).

There is, however, a deeper and even more inconvenient truth contained in his analogy. It is that the acceptance and use of these images has sanctioned, if not created, a potentially destructive theological climate. It is one where our vision of and relationships with God have been seriously distorted by looking at them through the lens of 'violence-of-God' traditions that supposedly separate out and favour some of us, and our interests, from those of the whole community of life on earth. This in turn has seriously distorted our vision of God's concern for all its other members in existence before and after our species' emergence on the planet.

Some present effects

Awareness of the effects of this supposed separation and consequent distortion of vision is now feeding into public consciousness. Commenting on the Audubon Society Report on the sharp decline in numbers of some of the most familiar and common birds in America, Verlyn Klinkenborg remarks that he has been reading such dire reports for many years now. They have the value, he says, of causing us to pay attention to species in trouble. But the sad fact is that the only species now likely to endure are the ones we humans manage to pay attention to (New York Times, 19 June 2007).

There was a time, he says, when it was better, if you were a nonhuman species, to be ignored by humans, because all the ones that unwittingly got our attention were trapped, shot or otherwise exploited. But in the past 40 years, we have killed millions of birds simply by going about our business as usual. The Audubon Society Report, he says, is really a report on who humans are: the unwitting and unconcerned destroyers of lives we do not even notice, let alone care for.

I absolutely agree with him when he says that our economic interests have proved to be completely incompatible with all but a very few forms of life, and that in our everyday economic behaviour we seem determined to attempt to discover whether we can live alone on earth. Yet the more we behave as if we are independent of the richness and diversity of biological life around us, the more we discover how dependent we are upon it. So, he says, environmentalists of every stripe argue that we must somehow begin to correlate our economic behaviour – every aspect of production, consumption, habitation – with the welfare of other species.

But the very foundation of our economic interests is self-interest and we see far too little self in other species to care much about their survival. So, Klinkenborg concludes, 'we look around us, expecting the rest of the world's occupants to adapt to the changes we have caused, when, in fact, we have the right to expect adaptation only from ourselves' (New York Times, 19 June 2007).

How are we to turn this around? When more than half the world's human population of almost seven billion are living in urban environments, how aware can we be that, as a species, we cannot live alone on earth? Are we sufficiently aware to stop separating our economic welfare (as usually understood) from the welfare or survival of other species never seen or taken account of in those environments? Do we believe that God separates their welfare from ours and sacrifices theirs to ensure our 'economic growth'?

Changing the ecotheological climate

A change in the theological climate will begin (but not end) by facing and refusing to support the prevailing understanding of ourselves and our individual interests as separate from those of other peoples and other species: that is, by re-envisioning our religious narratives within the larger context of earth's history. Discarding our theological partiality (in every sense) means giving up a view of ourselves as specially created and blessed by God, with its implication that we are outside general evolutionary processes and the constraints of ecological principles. Furthermore, we must learn to discard the idea that we are destined to enjoy a life with God in some distant heavenly world rather than the earthly one we now inhabit.

Setting aside this commonly held religious view of ourselves will help towards changing the behaviour that is its unintended consequence and will start to remedy the suffering that behaviour inflicts on humans and nonhumans alike. I suggest two allied approaches that may help us towards a different theological world view. The first is to see ourselves as emerging from the global environment and therefore as 'bodies-in-the-making', situated within the time frame, natural resources and spatial parameters of earth's history.

The second approach, bound closely to the first, is to look at the interrelational processes that bind us to more-than-human forms of life in other than economic terms. These processes are more accurately and fruitfully described as gift events, discerned and understood as such from a philosophical and religious perspective. This, as we shall see, involves more – far more – than seeing these events as an exchange of goods between two people. Ultimately it includes seeing interactions within the whole community of life on earth as an essential component of the self-making of all its members. Symbolized in terms of receiving and giving, the character and import of these interactions far exceeds any description of their features. The poet Brendan Kennelly aptly names this excess in these relational processes as 'the mystery of giving'.

Learning to see ourselves as situated within earth's history and participating within its relational processes will foster an ever-deepening understanding and awareness of our interdependent role and place within its life community. This may lead us – if not force us – to acknowledge the fact that, within that community, we are just another species; neither the owners nor the stewards of this planet (Lovelock 1995: 13). Like all other terrestrial species, we have always been, and still are, totally dependent on the life systems created and sustained by micro-organisms, life forms and living beings who came to be and to thrive over billions of years before our emergence. Up to now, this ancestry of life has made earth habitable for us. But climate change is teaching us that while our activities may, and do, destroy large portions of the fabric of life on the planet, we did not create it and, more importantly, we cannot re-create it the way it was. That is simply beyond human power.

Understanding and appreciating the coherence of our earth community relationships also requires changes in thinking about how we value ourselves. Presently, Western culture fosters a pursuit of individual self-determination, measured in personal, political, economic or academic achievement. And that in turn is measured in income and material assets won through competition rather than cooperation. This will change only as we come to understand the consequences for each of us of our dependence on the whole community of life, not simply for our personal development but, indeed, for our existence.

Such an understanding will become clearer as we explore the nature of events within the mystery of giving. It presents each of us as an outcome of this mystery, as a self that is created for a limited period through manifold, interdependent relationships and gift exchanges that are essentially community events. A crucial shift in our religious perspectives will follow from such a change in self-understanding. Our former belief that earth exists and was given to us by God solely 'for our use and benefit' will turn into a conviction that we would not, and cannot, exist without its gifts.

We might try to imagine what it would be like without plants that convert the sun's energy and regulate photosynthesis; without insects that pollinate plants; without birds or mammals or ants that move seeds around the ecosystems and without water that brings them to life; without micro-organisms in our gut that enable us to digest food. This will serve to increase our understanding of such serial and cyclic transformations and such symbiotic interdependence within and between micro and macro systems, and of their role in sustaining our lives. It will bring home to us how our survival depends on other life forms and natural life support systems, and help positively to shrink, not the earth, but the supposed distance between it and ourselves.

Re-envisioning theological relationships

This interplay between disciplines is central to developing an ecologically historic narrative. It will become a vehicle for change within theological narratives when we use it to elaborate explanations for and understandings of the relationship between God and the whole earth community. Understanding that relationship in terms of the mystery of giving is necessarily based on texts that do not sanction violence, in particular some of those from the Christian tradition centred on the life and teaching of Jesus. This will help us to adapt positively to the changes in world view brought about by climate change and allow nonviolent theological norms to emerge.

Such norms will be challenging. They will not support, for example, the traditional thinking behind a concept of salvation in which an angry, punitive God condemned Jesus to execution on a cross in order to be reconciled with sinful humanity. They will not allow the life and death of Jesus to be seen as God's way of effecting individual human redemption from eternal punishment for sin, or that such wording sums up the entire significance of Jesus and legitimizes forcible conversion to Christianity in his name.

Modern apocalyptic forms of such traditional thinking claim that the pain and suffering imposed by flooding, temperature rises, loss of energy sources, fresh water resources and desertification consequent on climate change – all primarily affecting the poorest members of societies – is divine punishment for sin. Or that it is the necessary, indeed longed for, preliminary to Jesus' second coming, when the world will end and God will punish sinners and reward the righteous. But more perceptive Christians will ask: what kind of God punishes sins perpetrated in Europe or in the United States by inflicting suffering on innocent communities in parts of the world remote from the perpetrators?

The mindset that presents this as divine punishment for human sin not only allows us to evade our responsibility for our contribution to its causes by lifestyle choices, our overuse of the earth's resources and our support for military power in all its aspects. It also exposes our failure to respond compassionately to those who suffer from the effects of such excessive use of power. So the poor women of Indonesia, Africa and Bangladesh continue to weep for themselves, for their children and for all those killed by famine, by waterborne diseases, by wars fought and fed by military-industrial-economic complexes.

This is the context out of which a re-envisioning of our relationship with God must emerge. It means, in Martin Buber's phrase, embodying it in the 'whole stuff of life'. That relationship cannot be 'preserved', he says, but only 'poured into life'; put to the proof in action in accord with each one's ability and the measure of each day (Buber 1970: 163). He expresses this relationship and its total demand succinctly:

> One cannot divide one's life between an actual relationship to God and an inactual I-IT relationship to the world – praying to God in truth and utilizing the world. Whoever knows the world as something to be utilized knows God the same way.
>
> (Buber 1970: 156)

Up to now our 'use' of the world, as demonstrated by colonial exploitation and the ravages of climate change, shows that we have seen it largely as a means of procuring national, individual, hereditary, monetary or spiritual gain. And it is a particularly blasphemous form of religious utilitarianism that uses violent and vindictive God traditions to legitimize our plundering the world's life resource bases for these ends.

For real religious relationships cannot be divorced from those that sustain our natural surroundings and connect us integrally to them. Such relationships are really one and the same thing considered from different perspectives. Implicitly and explicitly, studies of co-existence within the physical world can and do transform religious understandings of our role and place within the community of life on earth.

Within today's context such a transformation becomes apparent when we see ourselves not as owners of earth's resources but rather as participants in the mystery of continuously receiving and giving life within the earth community. Within a theological context such (r)evolutionary change takes shape when we approach God not as one who wields unaccountable power over earth and its inhabitants, but as one hidden in the mystery of creating and transforming all things. And so we may come to regard that mystery as a manifestation of the divine economy of love (Ephesians 3: 2, 9).

This early Christian expression of the mystery of giving not only offers an alternative 'economic' paradigm to that of the present destructive capitalist system that dominates, in every sense, the political, social and consumerist thinking of today. It also allows us to re-imagine the relationships between earth, ourselves and God as continually created and transformed by the mystery of giving. These re-imaginings portray a set of relationships based not on fear, aggression and exploitation of other living beings, but on love, freedom, nonviolence, forgiveness and generosity of spirit.

For some Christians, including myself, this divine economy may be symbolized in images given us through and in the historical Jesus. Existing, as we all do, on the boundary between human and divine, he is both described as, and expresses the gift of, God's love for the world (John 3: 16). His birth, life, teachings and death can then be perceived as having the character of a gift event, as a transformative, nonviolent, weak and therefore vulnerable force that, mysteriously, exceeds the power of the present world economy.

In his magisterial studies of the Sayings Gospel Q that he identifies as the earliest record of Jesus' own life and teachings, biblical scholar James Robinson remarks that a Jewish term for God acting on earth, used in Q, is God's *Wisdom*. In Greek *Sophia*, and in Hebrew *Hochma*, these names indicate the feminine dimension of God. Just as she had sent the prophets of the Old Testament (Q 11: 49–51; 13: 34–5) so she sent not only Jesus, but also John the Baptist. 'Wisdom was vindicated by her children' (Q 7: 35). In Luke's Gospel Jesus laments over the violence inflicted by Jerusalem on them:

> How often I wanted to gather your children together, as a hen gathers her nestlings under her wings, and you were not willing!
>
> (Luke 13: 34–5)

Robinson attributes this lament to Wisdom in all her manifestations and not exclusively to Jesus (Robinson 2007: xi). An early European religious culture would have attributed it to Gaia, an ancient Greek earth deity and female personification of the planet itself. Now resurrected scientifically as the name of James Lovelock's theory of the earth as a life-supporting planetary body, it is surely a contemporary sign of Gaian wisdom to lament the realities and effects of human-induced climate change – and to seek to remedy them.

4 The Second Historic Event

The second of Arendt's 'great events' that again confronts us with the phenomenon of alienation from earth is the Reformation. As this is an altogether
different kind of event, the alienation associated with it differs from that discerned in the discovery and appropriation of America. It is similar in that it
emerges from a distancing of ourselves from earth, but it diverges from it in its
effects. This is because the alienation arises not from a physical distancing of ourselves from the earth but by reducing our relationship with it to a purely
utilitarian one. Earth is viewed as something to be used, primarily as a means to
our salvation in whatever form this is understood. Therefore it is regarded as
something to be 'worked' by us, with that work done chiefly for the benefit of
our soul and secondarily, for profit.

The term 'Reformation' refers to a series of events and their effects, situated
within Europe, that reformed and reorganized the Western Church at the local,
regional and national levels during the sixteenth and seventeenth centuries in
ways that continue to affect us socially, economically and ecologically as well as
theologically. While essentially a religious and theological phenomenon, it was
strongly influenced by and, in turn, had a great influence upon wider cultural,
social, economic and political developments from the late medieval and early
modern period onwards (Bagchi 2000: 462–6). All of this testifies to the distinctions made between the name of an event, the time and place at which it occurs
and the excess or supplement potentially present in it.

A good example of the last is the way in which the potential within the
Reformation was released by access to the Bible in vernacular translations of
Scripture – such as Luther's – becoming available to lay people through that
great innovation, the printing press. This led to an extraordinary revival in
preaching and a radical empowerment of the laity, for it provided 'even tailors
and cobblers, even women and other simple folk' with an unchallengeable
authority to back up their insights and awkward questions. The biblical text was
read side by side with and through the lens of day-to-day happenings. At one
level it was a book like any other, full of stories and images to stir the heart. But
it was also an 'inner book', its contents being interiorized and collated by believ-
ers into their own personal rules of conduct. In the words of Tyndale, the
greatest of the early English translators, the only way to 'understand the scripture

unto our salvation' was to search it for God's personal covenant with us in baptism (Matheson 2006: 69–84).

Of particular relevance to climate change are the ways in which the Reformation's shaping of political and economic structures contributed to what is now called global capitalism, to its alienating effects on our relationship with earth, and our increased contribution to carbon emissions through industrialization and an expansion of resource consumption. Arendt quotes Weber's identification of the source of this type of alienation as an 'innerworldly asceticism' that itself is the 'innermost spring' of the capitalist mentality (Arendt 1958: 251–2). However, translating the religious phenomenon Weber termed *inner-weltliche Askese* as 'innerworldly asceticism' can be misleading.

In-the-world asceticism

For Weber it meant asceticism practised *within* the world: what today might be called '*in-the-world* asceticism'. He contrasted this with a monastic withdrawal from the world or 'other-worldly asceticism' (*ausserweltliche Askese*). As such, Weber said, there is a connection between *in-the-world* asceticism and the fundamental religious ideas of ascetic Protestantism as a whole, including its maxims for everyday economic conduct. He associated this type of asceticism with an English Puritanism derived from Calvinism and with Continental movements such as German Pietism (Weber 1930: 102, 140).

Weber was also punctilious – as I would wish to be – in stating that he was merely attempting to clarify the part that religious forces have played in forming the developing web of our specifically non-religious worldly modern culture. For this culture has resulted from complex interactions between innumerable different historical factors and events. His particular interest was to assess the extent to which certain characteristic features of this culture can be imputed to the influence of the Reformation. At the same time, he resisted the idea that it is possible to deduce the Reformation, as a historically necessary result, from specific economic changes. Countless historic events that cannot be reduced to any economic law had to concur in order that the newly emerging churches should survive at all. Nor does he hold (though this is often attributed to him) that capitalism as an economic system is a creation of the Reformation. His declared aim is to investigate whether, and at what point, certain correlations between forms of religious belief and practical ethics can be discerned (Weber 1930: 48–50).

Nearly a century after he wrote, in view of what has been said about the nature of events and in the light of what I shall say later about the effects of the Reformation on biblical scholarship, neither do I wish to appear to confine the emergence and effects of the Reformation within a particular time frame, or its vitality and potentiality to any particular effect. Nevertheless, I am grateful for his expert guidance in clarifying the manner and general *direction* in which Christian religious beliefs and practices formulated from the sixteenth century onwards have influenced our materialist culture.

However, I go beyond him in extending my concerns about their influence on human societies to how our relationship with the earth came to be perceived largely as one of independence from it. Bearing this in mind, I shall summarise and comment on some of the connections made by Weber between the fundamental religious ideas of ascetic Protestantism and their past and present effects on everyday economic conduct.

Weber made these connections largely by examining the writings of Church clergy. For this was a time when

> [T]he *beyond* meant everything, when the social position of the Christian depended upon his admission to the Communion and the clergyman, through his ministry, Church discipline and preaching, exercised an influence that we modern men are entirely unable to picture.
>
> (Weber 1930: 102) [Italics mine]

It is even more true that we in Europe today are entirely unable to comprehend such dependence on the clergy. It can still, however, be paralleled to some extent in parts of the United States and in other seventeenth-century European colonies served by members of churches and missionary societies founded at that time. Nor can the practical effects of this influence be discounted now, for better or worse, in religious attitudes to such issues as climate change. Historical and natural events were seen by Puritan eighteenth-century New England clerics like Jonathan Edwards to be communications from God. Earthquakes, military defeats or other disasters were often interpreted as calls to the elect to repent and to renew their commitment to God. They were part of God's great work of redemption in which all creation was involved and, though resisted by Satan's powers and the unrighteous, the final victory of good over evil was never in doubt (Schweitzer 2007: 246). Present-day echoes of this religious rhetoric resound throughout the United States whenever practical policies for reducing energy consumption are advocated.

Part of the reason for this can be attributed to the ambivalence of seventeenth-century Reformed clergy's discussion of wealth and its acquisition, in spite of an emphasis placed on New Testament teachings about the dangers of wealth and its pursuit. That pursuit is not only deemed senseless as compared with the dominating importance of the kingdom of God, but is morally suspect. This contrasts with Calvin's attitude to the acquisition of earthly goods. He saw them as no hindrance to the effectiveness of the clergy but rather as a thoroughly desirable enhancement of their prestige. Concerned about the survival of Geneva's economy (including its publishing business, so vital to his programme of reform) he endorsed the charging of interest on loans and the relaxation of other strictures on commerce, supporting challenges to the monopoly exercised over economic life by vested interests allied with the medieval Church (Compier 2007: 215–16).

Nevertheless, examples of condemnation of the pursuit of money and goods abound within Puritan writings. The true ethical significance of this condemnation, however, was a moral objection to 'the security of possessions' as a distraction from the pursuit of a righteous life and its reward in the 'beyond'. If we had everything we could have in this world, would that be all we hoped for? Complete satisfaction of desires is not attainable on earth because God has so decreed. Everlasting rest and satisfaction is in the next world. On earth we must 'do the works of him who sent us'.

> The span of human life is infinitely short and precious to make sure of one's own election ... This view does not yet hold, with Franklin, that time is money; but the proposition is true in a certain spiritual sense. It is infinitely valuable because every hour lost to labour is lost to labour for the glory of God.
>
> (Weber 1930: 104)

Work and profit

This emphasis on hard work, on continuous bodily or mental labour, was driven by two different motives. First, work has always been an approved ascetic technique or discipline within the Western Churches. But more importantly here, labour came to be considered *in itself* as the purpose of life, ordained as such by God, rather than, as Thomas Aquinas taught, being necessary for the maintenance of the individual and the community. It is true that there is a Puritan tendency to pragmatic interpretations in which the purpose of the division of labour is to be known by its fruits. The specialization of occupations leads to a quantitative and qualitative improvement in production and this serves the common good. But a change of calling is measured primarily in moral terms, that is, that a calling more pleasing to God is, as a general principle, a more useful one. We are called, chosen, elected by God to work in particular occupations.

Second, and in practice a most important motivation for an occupation, is that found in private profitableness. For if God shows one of his elect a chance of profit, he must do so with a purpose. Hence the faithful Christian must follow the call by taking advantage of the opportunity to get rich. Wealth is thus bad ethically *only insofar* as it is a temptation to idleness and sinful enjoyment of life. But as a performance of duty within a calling, its accumulation is not only morally permissible, but actually enjoined. This providential interpretation of profit-making justified the activities of the businessman (Weber 1930: 105–8).

Therefore, Weber says, the Puritan idea of being called, being chosen by God, together with the premium it placed upon ascetic conduct or work, was bound directly to influence the development of a capitalist way of life.

> As we have seen, this asceticism turned with all its force against one thing: the spontaneous enjoyment of life and all it had to offer.
>
> (Weber 1930: 111–12)

This includes the spontaneous enjoyment of the beauties and bodily experiences offered by the natural world. Weber cites as an example the fanatical opposition of the English Puritans to the royal ordinances permitting certain popular sports on Sundays outside of Church hours. A friend brought up in Scotland in a Free Presbyterian household in the 1920s went to school as usual on Christmas Day. It is not, Weber says, that the ideals of Puritanism implied a solemn, narrow-minded contempt of culture. Apart from a hatred of scholasticism, the great men of the Puritan movement were steeped in the culture of the Renaissance and perhaps no country was ever so full of graduates as New England in the first generation of its existence. But the theatre was obnoxious to them, as were Christmas festivities and everything that, to them, smacked of superstition.

While this worldly Protestant asceticism acted powerfully against the spontaneous enjoyment of possessions, it had the psychological effect of freeing the acquisition of goods from the inhibitions of a more traditional ethics. It broke the taboo against acquisitiveness. It not only legalized it, but (in the sense discussed) looked upon it as directly willed by God. The campaign against the flesh and dependence on external things was expressly not a struggle against the rational acquisition of wealth but the irrational use of it. In contrast to the glitter and ostentation of feudal magnificence was set the sober simplicity and clean comfort of what today would be called a middle-class home. More importantly, the religious valuation of continuous systematic work in a worldly setting as the most evident proof of genuine faith must, says Weber, have been the most powerful conceivable lever for that attitude to life which we call the spirit of capitalism. It stood at the cradle of modern economic man (Weber 1930: 117).

We also find that the effects of Arendt's first historic event are the 'given' of this, her second one. And we now live with the givenness of both and their effects. The theological colonization begun in earnest by the Roman Catholic Church in the fifteenth century proceeded apace in the three following centuries, compounded by the Reformation being situated in Europe where violent hatred for, and enmity between, Catholic and Reformed Christians was fought out in religious wars. Their effects endure in Northern Ireland for all to see. On both sides biblical and theological sanctions were cited for their actions, including the expropriation of lands, lives and bodies. This pattern was repeated with the colonizing and enslavement of non-Christian peoples, the appropriation of the products of their environments and the exploitation of their labour for monetary profit. The effects of that pattern are evident now in desertification, deforestation and biodiversity loss.

What we can also see is the way in which both despair and hope are religiously discerned and expressed in reaction to these events by reformers such as John Wesley. As Jennings says of him, he knew how to call upon those who profit from systems of exploitation to turn away from injustice in order to avert the judgment of God. But he also saw that more than self-interest (or soul interest) was at stake in their pursuit of profit at any cost. It was a betrayal of the ideals bequeathed to us and lived out by Jesus. And although he did not have the benefit of our postcolonial analyses of imperial policies, he does, as we shall see, offer valuable critiques and theological perspectives on the legacy of those policies (Weber 1930: 118–19).

The legacy of most seventeenth-century Christian religious reformers to their utilitarian successors was, however, above all, an amazingly good conscience about the acquisition of money so long as it took place legally. With the consciousness of standing in the fullness of God's grace and being visibly blessed by him, the bourgeois business man could follow his pecuniary interests and feel that he was fulfilling a duty in so doing. In addition, the power of religious asceticism provided him also with sober, conscientious and unusually industrious workers who clung to their work as to a life-purpose willed by God.

Finally, it gave him the comforting assurance that the unequal distribution of the goods of this world was a special dispensation of Divine Providence. Calvin made the much-quoted statement that only when the people, that is, the mass of labourers and craftsmen, were poor did they remain obedient to God. This idea was implicit in the justification of profiting from the slave trade and the keeping of slaves by such exemplary Christians as Jonathan Edwards. In its secular form it entered into later and current theories of the productivity of low wages.

By the twentieth century, Weber could see some of the historic effects of this seventeenth-century religious bequest. The essential elements of the spirit of capitalism were still the same, but without their religious basis. The Puritans wanted to work in a calling. We are forced to do so in the modern economic order. This is now bound to the technical conditions of mass production that, Weber said prophetically, determine the lives of all individuals '*and perhaps will so determine them until the last ton of fossilized coal is burnt*' (Weber 1930: 120–3) [Italics mine].

The development of property

The greatness of Weber's discovery about the origins of capitalism, says Arendt, lay precisely in his demonstration that an enormous, strictly mundane activity is possible without any care for or enjoyment of the world whatever. On the contrary, its deepest motivation is worry about and care of the self. This world alienation, she says, has been the hallmark of the modern age. Expropriation of monastic settlements by the Reformers – and before and afterwards, the Christian appropriation of countries, continents and peoples – meant the deprivation for certain groups of their place in the world and their naked exposure to the exigencies of life. This created the political conditions for the accumulation of wealth and the possibility of transforming that wealth into capital through labour.

Though she refers here to the event of the Reformation as the 'taking possession' and expropriation of ecclesiastical possessions, in her own context (and now in ours) it is impossible to read her words without remembering what this meant in her lifetime for the Jews and other non-ecclesiastical communities within Europe during Nazism. As Simone Weil observed, Hitlerism consisted in the application by Germany to the European continent, and more generally to the countries belonging to the white race, of colonial methods of conquest and domination (Weil 2003: 110). And indeed, of slave labour as a source of profit. '*Arbeit macht frei.*'

In regard to the analogous effects of the Reformation, Arendt wrote:

This propelled western society into a type of development where all property is destroyed in the process of its appropriation, all things devoured in the process of their production and the stability of the world undermined in a constant process of change.

(Arendt 1958: 255)

What distinguished this development from similar occurrences in the past is that expropriation and wealth accumulation did not simply result in new or expanded property or lead to a new redistribution of wealth, but were fed back into the process to generate further expropriations, greater productivity and more appropriation. This process, which has become the 'life process' of civil society, remains bound to the principle of world alienation from which it sprang. For the process can continue only if no consideration of the durability and stability of the earth is allowed to interfere with it; only so long as all earthly things, all end products of the production process are fed back into it at ever-increasing speed.

In other words, the process of wealth accumulation as we practise it – stimulated by the lifestyles and consumerism of civil society and in turn shaping human relationships through the manufacture of desire to consume – is possible only if the earth and the very earthliness of humans are sacrificed. And this process of alienation from the earth, started by expropriation and characterized by an ever-increasing growth in monetary wealth, assumes even more radical proportions if it is permitted to follow its own inherent law (Arendt 1958: 256–7).

Today, the capitalist process of wealth accumulation and private as well as national or regional appropriation of land is legalized through the historic evolution and use of two ancient institutions: property and money. They arose some time in the late eighth century BCE, gaining ground in Greece and in the whole of the ancient Near East. The most widespread assumption about their emergence is that money arose from bartering and was at the same time the means of increasing property ownership. Aristotle distinguished between two types of economy: one supplying households and the broader community with the goods needed to meet basic needs (*oikonomike*); the other used to increase monetary property for its own sake, buying and selling as part of an artificial form of acquisition (*khremastike*) (Duchrow and Hinkelammert 2004: 5). These distinctions become increasingly important, as we shall see, especially as the first is gradually subsumed into the second.

The connections between religious institutions – both physical and spiritual – and the ownership and management of property with the consequent accumulation of capital highlighted by Weber are also very early. An alternative economic theory to that of Aristotle connects the origin of money with the sacrificial practice of the temple and the collection of tributes. According to Deuteronomy (14: 24f.), in the late seventh century BCE the Judeans who lived further away from the central temple were supposed to exchange their sacrificial gifts for silver with which to buy sacrificial animals at the temple in Jerusalem. Evidence of small sacrificial spears as early forms of money reinforces this theory and also leads to the conclusion that the priesthood made a good profit from this business. In Jesus' later criticism of the temple (Mark 11: 15ff.) it becomes clear that the sacrificial system introduced under Solomon was a way of robbing the poor. This sacrificial function of money becomes all the clearer in its use within empires as a means to raise tribute from subjugated peoples (Duchrow and Hinkelammert 2004: 6).

John Locke and colonial property policies

In 1689, English Puritan John Locke (1632–1704) published his legitimization of property-owning in America as part of his endeavour to justify the colonialist policies of his patron, Lord Shaftesbury. He presented them metaphorically, as a political Genesis: 'Thus in the beginning all the World was America.' America was the second Garden of Eden, a new beginning for England if it could defend its claims to the continent against those of the Indians [*sic*] and other European powers. Like the world in the biblical Genesis, it is England's second chance at paradise, providing its colonial masters with a land full of all the promise of that first idyllic state. America represented a two-sided Genesis, a place to find and reconfigure both the origins of England's past and the promise of its future.

Like other writers defending England's right to American soil, he uses Genesis 1: 28:

> And God blessed them and said to them, 'Be fruitful and multiply, and fill the earth and subdue it; and have dominion over the fish of the sea and over the birds of the air and over every living thing that moves upon the earth.

Inevitably, all of them make the leap from subduing the land to claiming dominion. God, consequently, commanded the English not only to cultivate the land but to appropriate and hold dominion over it also. Barbara Arneil quotes Locke's conclusion:

> Hence subduing or cultivating the Earth, and having Dominion, are joyned together. The one gave title to the other. So that God, by commanding to subdue, gave authority so far to appropriate.
>
> (Arneil 1996: 140)

The term 'enclosure' is used repeatedly by Locke to define an individual's labour on the land to beget property and prevent others encroaching on it. In contrast to the Amerindians, this depended on the surveying and marking out of boundaries to individual pieces of property. Often they did not do this and only complied with English views when so commanded by English courts. Therefore they were judged, by Locke and others, to have no claim to property, as it is the act of enclosure along with that of cultivation that brings value to the land and so, if cultivated in common, it cannot be considered of any value. Therefore, the argument ran, the Amerindians who engaged in collective agricultural activities had no exclusive right to their land as property. Locke often refers to land in America as lying 'waste' (Arneil 1996: 141).

This concept of *vacuum domicillium* or 'waste land' corresponds to that of *terra nullius,* 'ownerless or empty land', used to deprive the Aboriginal Australians of their land rights up until the twentieth century. Again, this claim was based on the English perception of land not being 'used' to produce food or profit. But Locke's concern was not the self-interest of the individual owner. Like John Evelyn and the New England forests, he sees industry in America bringing wealth and fame to England as a whole. Only when you have access to both money and commerce with other parts of the world can the colonists go beyond 'mere subsistence' (Arneil 1996: 145).

A very important factor in Locke's theories was the role of money in transcending the limitations on expansion imposed by waste of land resources. He stresses its importance as:

> [s]ome lasting thing that Men might keep without spoiling, and that by mutual consent Men would take in exchange for the truly useful, but perishable Supports of Life.
>
> (Arneil 1996: 146)

God gave the land to be used by industrious and rational men. However, it was the ability to exchange the potential wealth of the land for hard currency that fuelled the massive appropriation of land by English colonists in the seventeenth century. Arneil quotes one of his unpublished notes on the necessity of trading to become rich:

> The chief end of trade is riches and power ... riches consist in plenty of movables that will yield a price to a foreigner ... especially in plenty of gold and silver.
>
> (Arneil 1996: 147)

With this 'invention of money', Locke observed, begins a state of affairs where those with money have a *right* to greater possessions. Not only that, the agreement to use gold and silver makes plain that 'Men have agreed to disproportionate and unequal possession of *the Earth'.* Consequently the right to property in land, initially begun by labour and limited by the waste produced, is extended only to those who have consented to the use of money. They are 'the Industrious and Rational' sons of Adam who have been given the 'Lord's garden' to be tilled and improved by them. Thus, while Englishmen in their own country or parish may agree through contracts to hold property in common and exclude all others from its use, Amerindians have no such rights. Their land exists under the 'law' for 'appropriating as commanded by God' (Arneil 1996: 148, 155).

Then and now, this exchange of land for money justifies the accumulation of as much land as possible in the same hands. It also marks the important move from money as payment in exchange for a product to money itself being a product. For while the indigenous inhabitants that do not use money are bound to keep to the natural boundaries of land they can use (*oikonomike*), the colonizers are not so bound, for they have agreed among themselves to the use of money (*khremastike*) (Duchrow and Hinkelammert 2004: 43–63). All that has been said here of land and its products applied also to its native inhabitants. It is no surprise that Locke's personal fortune was invested in the lucrative slave trade.

Theology and Locke's economic theory

In Locke's eyes, and in those of his Puritan readers, the theological legitimation for this colonization of the 'other' was his understanding of an all-powerful God. Since God, the source and origin of all power, can be seen as a type of owner of all creation, so too are humans in relation to the rest of the world. Whitney Bauman quotes Locke:

> God, who has given the world to men in common, hath also given them reason to make use of it to the best advantage of Life and convenience ... Whatsoever then he removes out of the State that Nature hath provided, and left it in, he hath mixed his *Labour* with and joined to it something that is his own, and thereby made it his *Property*.

Bauman goes on to ask: what is the 'state of nature' according to Locke? Theologically, it is the state of being fallen due to human sin. So nature was worthless until we acted to transform it into a new paradise. Also, our power to act mimics God's power, and our likeness to God consists in sovereignty over earthly existence and indeed, as Locke himself emphasized, extends to men having 'unequal and disproportionate possession *of Earth*'. As God creates out of nothing, so the human creates his individual property out of nothing – in the sense that Locke considered raw uncultivated matter to be awaiting human agency to give it value. Just as God is the sole source of value in creation, so humans made in the image of God bring value to valueless matter. So, Bauman notes, just as the chaos of Genesis 1: 2 gets erased in theological exegesis, so the prior presence of other non-Christian non-European humans and nonhuman others in the land gets erased (Bauman 2007: 360–1).

This is the immediate context for the Methodism that emerged in eighteenth-century England before the start of the Industrial Revolution. It can be seen as an attempt to address the problems of the secularizing influence of wealth, just as the earlier history of monasticism can be seen in the same way. John Wesley (1703–1791) understood the paradoxical relationships with riches at the heart of the Puritan reforms. As riches increase, so will pride, anger and love of the world in all its branches. As Methodists grow diligent and frugal, so their goods increase, and there is a proportionate increase in pride, in anger, in the desires of the flesh. So, although the form of religion remains, the spirit is swiftly vanishing away. Is there no way to prevent this?

British colonization of North America was the context for Wesley's vision of 'the community of goods' prescribed in Acts 2 and 4 and was at the heart of his desire to go and minister in Georgia in 1735. For him that vision would be realized by living and working among the Native Americans, where a simplicity of lifestyle was founded on what he called 'the pre-eminence of the Heathen'. He accorded this eminence to them because they desire and work for nothing more than plain food to eat and plain raiment to put on and seek this only from day to day. They lay up for themselves no treasures upon earth, no stores of purple, linen, gold or silver.

> But how do Christians observe what they profess to receive as a command of the most high God? Not at all! Not in any degree; no more than if no such command had ever been given to man.
>
> (Jennings 2007: 259)

Later in Wesley's career his passion for this vision focused his attention on the two main pillars of the emerging British Empire and the sources of its wealth: the slave trade and the franchising of India under the East India Trading Company. This marks an important interaction and confluence between Arendt's first and second historic events. One of Wesley's arguments against slavery was his insistence on the virtues of those societies from which slaves were taken. Noting the remarkably just customs of their people and of their general regard for the poor and the ill, he concludes:

> The Negroes who inhabit the coast of Africa, from Senegal to Angola, are so far from being the stupid, senseless, brutish, lazy barbarians, the cruel, fierce perfidious savages they have been described as: on the contrary, they are ... remarkably sensible ... industrious to the highest degree ... fair, just and honest in all their dealings except where white men have taught them to be otherwise ... Where shall we find at this day, among the fair-faced natives of Europe, a nation generally practising the justice, mercy and truth generally found among these poor Africans?
>
> (Jennings 2007: 260)

Justice, mercy and truth are the very characteristics that Wesley often associates with the renewed image of God. His religiously based appreciation of the virtues of Native American and African cultures undercut what would become one of the principal legitimations of colonization and emergent empire: the civilizing mission of Europe. In today's new form of empire, remarks Theodore Jennings, this rationale has become the mission of extending freedom (of the financial markets) and 'democracy' (as the politics of marketing) to other nations and cultures. Wesley's appreciation of the virtues of other cultures, combined with his keen empathy for those who were the victims of conquest, avarice and violence, led him to issue the following broadside against the imperial designs of the great nations:

> It were to be wished that none but heathen had practiced such gross, palpable works of the devil. But we dare not say so. Even in cruelty and bloodshed, how little have the Christian nations come behind! And not the

Spaniards or Portuguese alone, butchering thousands in South America. Not the Dutch only in the East Indies, or the French in North America, following the Spaniards step by step. Our own countrymen too have wallowed in blood and exterminated whole nations: plainly proving thereby what spirit it is that dwells and worked in the children of disobedience.

(Jennings 2007: 261)

His greatest concern is the spirit behind the emerging imperial designs of England. In the same sermon he points to the suffering of the peoples of India and Africa in order to indicate the limits to our understanding of God's providential ruling over history. 'How many hundred thousands of the poor quiet people have been destroyed and their carcasses left as dung of the earth!' In such texts we see a Wesley who has moved from being an observer of human misery, who simply must have recourse to the inscrutable providence of God, to one who finds the causes of human misery in the imperial policies of his own nation. Jennings remarks: 'One might say that his general orientation has changed from that of a Roman stoic to that of a Hebrew prophet' (Jennings 2007: 257–64).

Locke's enduring legacy

The 'settling' of the right to property by virtue of 'best use' is no longer based openly on Locke's theological distinctions. Yet their effects have been universal in scope. Used originally to justify colonial expansion they underlie and underpin today's multinational market trading and globalizing capitalism.

This has had massive implications for, and input into, climate change. For the global market depends on property used for the accumulation of wealth (founded on an agreement on the use of money) in contrast to property serving to guarantee the needs of life (property in the state of nature). Capitalist property is distinguished by use and monetary value in order to devalue property 'in the state of nature'. It is this that has created and sustained our contributions to climate change and that makes it very difficult for us to change our attitude to the land and therefore to reduce those contributions.

The concept of 'capitalist property', given absolute preference by Locke as an argument in favour of the conquest of North America, continues to be used as an argument to justify expropriation and exploitation for profit of the natural resources on which all life depends. Then, it meant that the colonizer could exploit 'the state of nature' in order to occupy far-off lands without being bound by the limits in land use 'for sustenance' that Locke set the indigenous population. If they accepted money for their land, they gave 'tacit and voluntary consent' to the system and lost their right to the land. If they did not accept money, they also lost their rights. The only alternative they were offered was to be shot or hanged.

The conclusion of this process is that humankind appears to consider itself constituted by property ownership. Individuals appear to share in humanity through being owners and have dignity only in so far as they are owners. Anyone who doubts this conclusion need only look at the growth in home ownership in Britain and the United States over the past century and the resultant chaos in current world markets caused by the collapse of confidence in sub-prime mortgages (Duchrow and Hinkelammert 2004: 59–67).

Present effects of appropriating the earth

In fairness to Locke, he did recognize a category of 'property in the state of nature serving to guarantee the needs of life' – albeit solely human life – as well as 'property for the accumulation of wealth', that is, of money. Today, however, the first category has almost disappeared from view as a result of industrialization, growth in mass production and increasing urbanization. However, there is hope in the fact that Locke's definition of the 'state of nature' has been translated into the concept of the 'commons' and indeed into movements in the United States where it is being used to 'reclaim' them for the common good. This is in contrast to present practice where

> corporations take valuable stuff from the commons and privatize it. With the other hand, they dump bad stuff into the commons and pay nothing.
>
> (Barnes 2006: 16–20)

The reason for this is that, as Hinkelammert demonstrated in an earlier work, the *spirit* of capitalism now works in two apparently opposing ways. First, the language of commodity relations is used to make it seem as though the relationships between human beings and the effects of the division of labour on human life are two totally independent and unconnected issues. As I understand it, this means that commodity relations – between a commodity and its producer, between the one who determines the market price and sells it and the one who buys it, and between all of these and those who fix its share price and gamble on it in the commodities market – rely on a tacit and therefore impenetrable division between the commodity produced and the situation of the individual who produces it, and between that individual and those others who, once it is produced, relate to it as though that individual does not exist. And supporting the whole process is the invisibility and consequent devaluation, literally and practically, of natural resources, life support systems, 'the state of nature' or 'the commons'.

Commodity markets rely on the invisibility of workers. So the relationships between the commodity's 'future' on the stock market and the workers' future also remain invisible. The workers themselves, as individuals, disappear not only in the eyes of those who employ them, but in their own. They see themselves as slaves to their machines and to the commodities they produce. They do not 'count': only commodities do. While on the one hand commodities are objects, on the other hand they are given the dimension of players in the economic process. Commodities-as-subjects are described and set up as having relationships among themselves. Coffee has a 'future', as does coal, iron and steel. Oil 'fights with coal for market share' and so on (Hinkelammert 1986: 6–29).

Since Hinkelammert wrote this, growing awareness of these commodity relations and their effects has given rise to a variety of non-governmental movements focused on the commons or on its products and their production that, like Fairtrade, have had a limited, albeit welcome, effect on the lives of commodity producers. Now, in a time of climate change, we are also increasingly aware that there is another 'invisible' player involved in these commodity relations: the earth itself. What Locke called 'the property in the state of nature serving to guarantee the needs of life' corresponds to what we call 'natural life support systems'. As we come to better understand them, we begin to realize that they cannot be regarded as anyone's 'property'. Such a change of viewpoint raises serious questions about production processes that require an increasing use of 'property in the state of nature' to maintain sustained growth in the accumulation of wealth. The currently accepted vector of growth, calculated as Gross Domestic Product (GDP), depends on an increased appropriation of vital life resources in land, water, raw materials and energy to continue the cycle of production, consumption (of products and resources) and waste generation.

In order to profit from this cycle, one must have money. Money is now itself a commodity that stands out above the rest because it serves as their common denominator; one into which the commodity, its producer and the workers who convert it into a commodity have their value assessed and confirmed. Therefore money has become *capital,* moving beyond Weber's important distinction between 'capital' and 'capitalism'.

Capital, he said, is the estimated value of the material means of production, such as raw materials, buildings and machinery, balanced against liabilities (Weber 1930: xxxii–xxxiii). Now, however, as Duchrow and Hinkelammert point out, as money itself has become capital, or 'property generating profit', it has also become programmed to multiply itself. Economic success (GDP) is now measured exclusively in terms of monetary growth. The fact that a rise in share prices can result from destructive elements such as major accidents that require additional services – the repair and building of flood defences, job redundancies, and so on – points to the inadequacies of such a measure. Economic activity should not be seen purely from the angle of wealth accumulation but in terms of its effects on people's lives and their environment. There have been attempts to introduce social and ecological indicators into the measurement of economic progress, the most well known being the Human Development Index of the United Nations Development Programme (Duchrow and Hinkelammert 2004: 187).

There is also, they say, an inbuilt injustice in the *added value* role of money in the property market economy. Countries with 'hard' currencies – in particular the USA with the dollar – can trade and go into debt in their own currency. Yet the currencies of most of the postcolonial countries are not accepted in international trade and debt management. They must therefore earn hard currencies through exports or take out loans – with the consequent unrepayable debt. This has led to the scandal of Zambia's debt, for example, being sold off as a commodity, with the purchaser being able to sue for massive repayments, including accumulated interest, in the English courts.

However, there is another disquieting dimension to money and debt being used as a commodity to be bought and sold worldwide. Who actually owns the money in circulation? Most of us think it is the state, because the issuing of coins and notes is a state 'privilege'. In reality, the state now contributes only about 3 per cent of the money supply in circulation. The rest is obtained in different ways from private financial institutions, primarily banks. For instance – and this is a very real concern at the moment – the banks issue money as a loan in exchange for a mortgage on a home. The same applies to land, industry, public institutions and services and, above all, to the state budget itself.

Consequently banks make money on the basis of the debts of citizens and the public sector. This 'debt-money' is owned and controlled by the banking system and entails two decisive disadvantages for citizens and the public as a whole. First, their property moves increasingly into the hands of capital owners and their agents, the banks. Second, the money used to repay debts always involves inter-est payments to the capital owners. If the banks grant loans and thereby make money (on the basis of only a small share of their own capital), repayment means the debt is cancelled but not the amount created by making interest on the loan.

> This means that we, the political community, allow the capital owners, rep-resented by the banks, to create further property for themselves by making money on debts without doing a stroke of work.
>
> (Duchrow and Hinkelammert 2004: 188)

The Puritan ideal of labour as a calling from God has degenerated into that of a work-free, leisure-rich, high consumption lifestyle. Reading the financial papers today we can see some of its worldwide effects in the collapse of the 2007 sub-prime mortgage market in the United States. Sadly, just such a situation was foreseen by their third President, Thomas Jefferson (1801–9):

> If the American people ever allow the banks to control the issuance of their currency, first by inflation and then by deflation, the banks and the corpora-tions that will grow up around them will deprive the people of all property until their children will wake up homeless on the continent their fathers occupied. The issuing power of money should be taken from the banks and restored to congress and the people to whom it belongs. I sincerely believe the banking institutions having the issuing power of money are more dan-gerous to liberty than standing armies.
>
> (Rowbotham 1998: 34f.)

Responding to this situation

Duchrow and Hinkelammert give a valuable overview of various joint civilian and Church responses to situations where common environmental goods, such as water, are the subject of takeover bids by private corporations. In many instances these alliances are a response to particular events such as water privatization in

Bolivia and Brazil. In this they can be compared to the imposition of the Salt Tax in colonial India and Gandhi's response to it (Duchrow and Hinkelammert 2004: 172–7). In what Martin Luther King Jr called 'the fierce urgency of now', they have become a legitimate, indeed an imperative Church response to the effects of climate change on common environmental goods.

Duchrow compares this challenge for the Churches to that of Nazism and apartheid. He reprises Dietrich Bonhoeffer's argument for opposing the totalitarian claims of Nazism where there was 'too little' protection of some of its citizens, notably the Jews, the disabled and the homosexual, and 'too much' state influence when it declared itself absolute and thereby, an idol. This moment, he says, is a *status confessionis,* that is, a point at which the Church must make a clear decision and take a strong stand with all the consequences this entails. The confessors did so with the Barmen Declaration (1934), by which they constituted themselves as the 'Confessing Church'; in contrast with the German Evangelical or 'Reich' Church. This led to the question of how a 'Church Community' is constituted. Bonhoeffer did not begin by asking who belongs to the true Church. Rather the Church must wrestle with giving visible witness to Christ in real-life situations.

What does this mean in a time of climate change? It means, in the global context of this chapter, becoming aware of the present effects of the unbridled capitalist international system that promises wealth to the few while destroying the environmental capital that sustains the life of the whole earth community. It also calls for us to see how this intersects and interacts with the events chronicled in the previous chapter. It means acknowledging that the religious roots of this event can be traced, in some measure, to what we call the Reformation and that they depend on an image of God who calls an elect few to a way of life rewarded in another world and earned by the acquisition of monetary wealth in this one.

This has necessarily alienated us from earth, in that it is seen as a means to an end, the salvation of our souls. As a possession to be utilized in this way, all life on earth is devalued. At a time characterized by a desire for ever-increasing monetary wealth, this world view has contributed to the evolution of a consumerist culture and economic structure ruled by buying power, one where to be poor is to be a loser.

But it is also important to acknowledge that figures such as John Wesley, Bonhoeffer and the continuing witness of 'confessing' Churches and communities, of all denominations and none, are also part of the legacy of the Reformation. And as I shall show in later chapters, so too is the biblical scholarship that has given us insights into the life and teachings of Jesus that can usefully inform our theological reflections in this time of capitalist dominance and climate change. Together they give us hope that we might, after all, be able to fulfil Bonhoeffer's vision and make the loving care of God so believable that all human *economic activity* would accept it as a model (Duchrow and Hinkelammert 2004: 212).

Alfred North Whitehead considered the Reformation, for all its importance, to be a domestic affair of the European races viewed, even by the Christians of the

East, with profound disengagement. It was, he said, a transformation of religion but not the coming of religion. Nor did it claim to be. The Reformers maintained that they were only restoring what had been forgotten. But it is quite otherwise, he said, with the rise of modern science (Whitehead 1925: 1–2). With this in mind we will look at the third historic event Arendt notes as a force alienating us from the earth – the discovery of the telescope by Galileo.

5   The Third Historic Event

The previous chapter ended with the assertion by Whitehead that the discoveries of Galileo have had more impact than the Reformation in alienating us from the earth. Living with the effects of both these events for another eighty years after he made those comments, one must agree with him and with Arendt that in many ways Galileo's discovery has been the one with the greatest impact. She quotes Whitehead: 'Since a babe was born in a manger, it may be doubted whether so great a thing has happened with so little stir.' Nothing in these words, she says, is an exaggeration. Like the birth in a manger, which spelt not the end of antiquity but the beginning of something so unexpectedly and unpredictably new that neither hope nor fear could have anticipated it, those first glances into the universe through an instrument adjusted to human senses and destined to uncover what definitely lies beyond them, set the stage for an entirely new world view and determined the course of other events (Arendt 1958: 257–8).

It determined that today our view of the earth as 'a small blue planet' is a commonplace one. But there is nothing commonplace about the development of a science that made it possible for such an image now to be imprinted on our minds. For that science enabled us to consider the nature of the earth from the viewpoint of the universe. This is

> [T]he 'Archimedean point' that now epitomises the scientific point of view: one which, by using instruments adjusted to the human senses, uncovers what definitely and forever must lie beyond them.
>
> (Arendt 1958: 248)

Galileo's discovery and use of the telescope meant that the 'secrets' of the universe, until then hidden or conjectured, became accessible to human minds with the certainty of sense-perception. He put within the reach of us earth-bound creatures what seemed to lie forever beyond the grasp of our body-bound senses. With the invention of the microscope, this also became true for subvisible organisms on earth, and with the invention of the electron microscope the very molecules of life can be perceived as can subatomic particles in the universe.

Historically, his discovery provided the means for what Arendt saw as realizing both our worst fear and our most presumptuous hope. The fear is that our senses, the very organs we use to receive and to access reality, might betray us. The hope is to find the Archimedean point outside the earth from which to look down on her as if we were inhabitants of the sun. Both the fear and the hope emerged inseparable from this event.

At the time, however, except for a numerically small, politically inconsequential number of learned men – astronomers, philosophers, theologians – the telescope created no great excitement. Indeed, many of them rejected Galileo's offer of a chance to use it. This annoyed him greatly.

> His opponents' resistance appeared to him a symptom of an intellectual indolence that entrenched itself indoors in studies, to 'look things up' in Aristotle instead of looking at the things themselves. Galileo did not reflect on the complicatedness of natural optics, which does not automatically adjust to a technical increase in what it can accomplish.
>
> (Blumenberg 1987: 660)

If it was difficult to see what one does not expect – such as the moons of Jupiter – it was almost impossible to accept, as a result of mere optical experience, what is not admissible in the context of one's a priori understanding of the world. Blumenberg quotes Goethe's remark that one perceives only what one already knows and understands. 'Often one fails to see for many years what only a more mature state of knowledge and culture enables us to become aware of in an object that confronts us every day.' Blumenberg goes on to make an important point:

> The fool's role that Galileo's opponents have long played in the historiography of natural science has rendered them harmless for us and obscured their significance as indicators of the difficulties in our relation to reality that are always present and become especially acute in historical situations where radical change is under way.
>
> (Blumenberg 1987: 662)

Radical change in our own time

Today we ourselves face 'difficulties' in our relation to reality as just such a radical change is under way in 'an object that confronts us every day', that is, in our climate and in how its changing character affects and will increasingly affect our lives. It is certainly true that many of us, even in this age dominated by science, have difficulty in accepting the reality of that change presented to us in scientific formulae and argued for on the basis of experimental results published in research journals. Therefore, even as we are bombarded with more and more facts, images and figures about climate change, we find it almost impossible to accept our role in it and our responsibility for its effects. So we shirk changing the routine conduct that contributes to those effects.

Part of the problem is that, unlike Galileo's opponents, we *do* look at the 'things themselves': we are aware of and note the effects on our immediate environments of increased rainfall or lack of rain; of temperature rises; of changing migratory patterns and, at one or more remove for most of us, of glacier melt and sea rises. But we perceive in these phenomena only what we already know and understand about variations in weather patterns, their seasonal effects and recurrences.

So while we are offered, indeed exhorted, to adopt an Archimedean view of our planetary climate, the radical changes in it do not appear to match the reality conveyed by our senses. Therefore, being told about them provokes a mixture of fear and denial in most people because the gap between our experience of the weather and the reality of climate change is too great to be bridged by scientific authority alone. This reaction is intensified by a growing realization that the change in climate can only be dealt with by a change in lifestyle among the most affluent nations.

Looking back to Galileo, there had been religious and cultural antecedents for his astrophysical world view, notably in the philosophical and theological speculations of Nicholas of Cusa and Giordano Bruno about the (in)finite nature of the universe and in the mathematically expressed projections of Copernicus and Kepler. However, neither the speculations of philosophers nor the imaginings of astronomers has ever constituted an event. But after Galileo's findings were confirmed, there was a sudden change of mood in the learned world. The enthusiasm with which Bruno had conceived of an infinite universe, the pious exultation with which Kepler had contemplated the sun and the more sober speculations of Nicholas of Cusa were replaced by demonstrable facts – that there are eight, not seven planets and that the moon does not have a smooth and polished surface. The immediate and consequent philosophical reaction was not exultation but Cartesian doubt.

> For many centuries the consequences of this event, again not unlike the consequences of the Nativity, remained contradictory and inconclusive, and even today the conflict between the event itself and its almost immediate consequences is far from resolved.
>
> (Arendt 1958: 259–261)

Some consequences of this event

What kind of consequences are we talking about? First among them was the emergence and gradual development of what Arendt calls 'the modern astrophysical worldview' and its challenge to the adequacy of the senses to reveal reality. This has left us in a universe of whose qualities we know no more than the way they affect our measuring instruments. For – in Arthur Eddington's words – 'the former have as much resemblance to the latter as a telephone number has to a subscriber.' Heisenberg made the same point, says Arendt, by concluding that instead of nature or the universe, we encounter only ourselves. 'For the observed object has no existence independent of the observing subject.'

Schrödinger described what he called 'this very strange state of affairs':

[O]n the one hand all our knowledge about the world around us, both that gained in everyday life and that revealed in the most carefully planned and painstaking laboratory experiments, rests entirely on immediate sense perception; while on the other hand this knowledge fails to reveal the relations of the sense perceptions to the outside world, so that in the picture or model we form of the outside world, guided by our scientific discoveries, all sensual qualities are absent.

(Schrödinger 2000: 153)

It is as though the wish for an Archimedean point would be granted only provided we lost touch with earthly reality (Arendt 1958: 262).

With the development of the Hubble Space Telescope, we can now test the truth of this conclusion. A commentary by Anthony Doerr in *Orion* magazine on the NASA site, http://hubble.nasa.gov and http://www.hubblesite.org of the Hubble Ultra Deep Field, gives us the following descriptions of the earth.

We live on Earth. Earth is a clump of iron and magnesium and nickel, smeared with a thin layer of organic matter and sleeved in vapor. It whirls along in a nearly circular orbit around a minor star we call the sun ...

The Earth is massive enough to hold all of our cities and oceans and creatures in the sway of its gravity. And the sun is massive enough to hold the Earth in the sway of its gravity. But the sun itself is merely a mote in the sway of the gravity of the Milky Way, at the center of which is a vast, concentrated bar of stars, around which the sun swings (carrying along Earth, Mars, Jupiter, Saturn, etc.) every 230 million years or so. Our sun isn't anywhere near its center; it's way out on one of the galaxy's arms ...

As of early April 2007 astronomers had found 204 planets outside our solar system. Chances are many, many stars have planets or systems of planets swinging around them. What if most suns have solar systems? If our sun is one in 10 sextillion, could our Earth be one in 10 sextillion as well? Or the Earth might be one – just one, the only one, the one. Either way, the circumstances are mind-boggling.

Doerr concludes by asking his readers to take a moment every now and then 'to *leave the confines of ourselves* and fly off into the hugeness of the universe, to disappear into the inexplicable, the implacable, the reflection of that something our minds cannot grasp'. [Italics mine]

This exemplifies the alienating effects of what Arendt calls 'handling nature from a point in the universe outside the earth'. Nothing occurring on earth is viewed as an earthly happening for all earthly events are perceived as subject to universally valid laws in the fullest sense of the word. That means, among other things, that they are viewed as subject to laws valid beyond the reach of human sense experience. Therefore they are seen as being beyond the influence of human action. It is important to compare the effect of this world view with those based on 'earthly' chaos theory alluded to in the Introduction, and to remember Prigogine and Stengers' emphasis on that theory's implications for the significance of individual actions in a time of climate change.

Here, however, the 'implacable' workings of universal laws are presented to us in images from and of the Hubble Deep Field that are themselves based on computer calculations and are accessible to our eyesight only on a computer screen. The validity of these images lies far beyond the reach of human memory and the appearance of humankind on earth; beyond the coming into existence of organic life and even of earth herself. As Doerr acknowledges, they are the *reflection* of something our mind cannot grasp.

Therefore, they are *twice* removed from our grasp of reality and, by analogy, removed from any effect we might have on them – even though they are presumed to govern life on earth. The real appeal of 'leaving the confines of ourselves' and flying off into the hugeness of the universe is that, by doing so, we leave behind our sensory inputs and responses and with them, our responsibility for what happens here on earth.

The laws Doerr invokes are formulated from an Archimedean point that we have moved farther and farther away from earth: to a point where neither sun nor earth are centres of a universal system and we ourselves no longer feel bound even to the sun. Now we can establish ourselves as 'universal' beings, creatures who are terrestrial but who, by virtue of reasoning, can overcome this condition not merely speculatively but in actual fact. We are now promised tours in outer space; expeditions to the moon and to Mars.

From a closed world to an infinite universe

All this signals a theoretical and increasingly technologized shift not just from an earth-centred world but even from a heliocentric one. We have moved to a centreless world view: one where we assume we can 'fly off into the hugeness of the universe'. The scientific horizon has shifted decisively beyond the Copernican world view of a closed, finite world to an infinite universe. Copernicus believed in the existence of material planetary spheres because, says Alexander Koyré, he needed them in order to explain the motion of the planets. He also believed in a stellar sphere of fixed stars that he no longer needed. But even though its existence did not explain anything, it was useful in that it 'embraced and contained everything and itself'. It held the world together and enabled Copernicus to assign a determined position to the sun. Therefore the *world* of Copernicus remained a finite one, encompassed by a material sphere of the fixed stars with a centre occupied by the sun (Koyré 1957: 31–2).

Nevertheless, after Copernicus it was somewhat easier, psychologically if not logically, to pass from a very large immeasurable and ever-growing world to an infinite one. His reform of astronomy removed one of the strongest scientific objections to the infinity of the universe based on the motion of the celestial spheres. It is, therefore, not surprising that in rather a short time after him, some bold minds made the step he himself refused to make and asserted that the starry heavens, in which the stars are placed at different distances from the earth, 'extendeth itself infinitely up' (Koyré 1957: 35).

The first to do so, says Koyré, was Thomas Digges in 1576. However, it was Giordano Bruno, in his *La Cena de le Ceneri,* who gave the best pre-Galilean discussion and refutation of the classical objections – Aristotelian and Ptolemaic – against the motion of the earth. And to the old and famous question: why did not God create an infinite world? – one to which the medieval scholars gave such a good answer, namely, by denying the very possibility of an infinite creature – Bruno simply replied, and is the first to do so: God did. And even: God could not do otherwise. Bruno's God could not but explicate and express himself in an infinite, infinitely rich and infinitely extended world. Bruno expressed his vision of this world in almost rhapsodic terms (paralleled in today's enthusiastic comments on the vision opened to us by the Hubble Space Telescope):

> Thus is the excellence of God magnified and the greatness of his kingdom made manifest; he is glorified not in one, but in countless suns; not in a single earth, but in a thousand, I say, in an infinity of worlds.
>
> Thus not in vain the power of the intellect which ever seeketh, yea, and achieveth the addition of space to space, mass to mass, unity to unity, number to number, by the science that dischargeth us from the fetters of a most narrow kingdom and promoteth us to the freedom of a truly august realm, which freeth us from an imagined poverty and straineth to the possession of the myriad riches of so vast a space, of so worthy a field of so many cultivated worlds.
>
> (Koyré 1957: 32–42)

It has often rightly been pointed out, says Koyré, that the loss by the earth of its central, and thus unique, situation in the cosmos led inevitably to the loss, by man, of his unique and privileged position in the theo-cosmic drama of creation in which he was, until then, the central figure. However, as this 'loss' was gained through the use of human intelligence, it led instead to a new form of anthropocentrism that effectively made the evolution of human intelligence coincident with the evolution of the universe. Its formulation as 'the anthropic principle' (to which I shall return) expresses a deepening alienation of the human being from that 'most narrow kingdom', the earth, in favour of a non-geocentric, universal 'observership' status. With this emerged a rational-instrumental relationship with earth that corresponds in some measure to the 'use' of the world for personal salvation or profit discernible in the 'in-the-world' asceticism of the Reformers.

The language of universal science

How was this observership status attained? By employing the human imagination and mind in the concurrent development of modern geometry and algebra and in the emergence of calculus in the seventeenth century. The decisive point here is that without this non-spatial symbolic language Newton would not have been able to unite astronomy and physics into a single science or, to put it another way, to formulate a law of gravitation where the same equation, as we saw in the report from Hubble, is used to cover the movements of heavenly bodies and the motion of terrestrial bodies on earth.

By doing this, modern mathematics, in an already breathtaking development, has discovered the ability to grasp in symbols those dimensions and concepts that at most had been thought of as negations and therefore limitations of the mind. For their immensity not only *appears* to transcend minds whose existence lasts an insignificant length of time and remains bound to a relatively unimportant corner of the universe, it *does* transcend them.

Even more significant for Arendt was the fact that these new mental instruments opened the way for an altogether novel mode of meeting and approaching nature empirically. Through thought experiments, instead of observing natural phenomena as they occur, nature was placed under conditions won from and based on a universal, astrophysical viewpoint, one outside earthly nature itself. For this reason mathematics has become the leading science of the modern age and, with advanced computer technology, has become the 'eyes' of the mind just as bodily vision interprets sense data.

Mathematics has not simply enlarged its content or reached out into space to become applicable to the immensity of an infinite and infinitely expanding universe. It has ceased to be concerned at all with phenomena and has instead become the science of the structure of the human mind. Mathematicians have succeeded in reducing and translating all that is not man into patterns that are identical with human mental structures, eventually removing the eyes of the mind, no less than the eyes of the body, from the multitude and variety of all that constitutes our concrete experience. This has replaced the traditional scientific method of classification with measurement. The former follows objective realities whose principle is found in the otherness of nature: the latter is entirely subjective, independent of qualities and requires no more than that a multitude of objects be given (Arendt 1958: 264–8).

Foremost in Arendt's mind – she wrote during the Cold War – was the enormously increased human power of destruction.

> We are able to destroy all organic life on earth and shall probably be able one day to destroy even the earth itself. However, no less awesome and no less difficult to come to terms with is our corresponding new creative power; that we can produce new elements never found in nature, that we are able not only to speculate about the relationships between mass and energy but actually to transform mass into energy or to transform radiation into matter.
> (Arendt 1958: 269)

This echoes the reaction of Robert J. Oppenheimer, physicist and leader of the Manhattan Project, when the first nuclear fission bomb went off in the desert outside Alamogordo, New Mexico on 16 July 1945. As the colour of the morning sky changed abruptly from pale blue to blinding white, he felt at first a surge of elated reverence. Then a sombre phrase from the *Bhagavad Gita* flashed into his mind: 'I am become Death, the shatterer of worlds.' For the first time in some two million years of human history there existed a man-made force capable of destroying the entire fabric of life on the planet (Worster 1977: 339).

Arendt also noted that we had begun to populate the space around earth with man-made stars, or satellites. And that there was a hope that in the not too distant future we would be able to perform what before was regarded as the 'deepest, holiest secret of nature, to create or re-create the miracle of life'.

She deliberately used the word 'create', she said, to indicate that we are actually doing what all ages before ours thought to be the exclusive prerogative of the divine. Though this may strike us as blasphemous, it is no more blasphemous than what we have been and are aspiring to do. For the thought loses its blasphemous character as soon as we understand what Archimedes understood so well, even though he did not know how to reach his point 'outside' the earth. That is, that no matter how we explain the evolution of earth, nature and man, they must have come into being through some 'beyond-this-earth', transmundane force; one whose work must be comprehensible to the point of imitation by someone able to occupy the same location.

> Philosophically, it seems that man's ability to take this cosmic, universal standpoint without changing his location is the clearest possible indication of his universal origin, as it were. It is as though we no longer needed theology to tell us that man is not, cannot possibly be of this world even though he spends his life here.
>
> (Arendt 1958: 270)

The anthropic principle

Some further theological implications of this world view will be examined in later chapters. Here I shall take a closer look at the view of ourselves implicit in scientific developments by which, as Koyré remarked, man lost the world in which he lives and about which he thinks, replacing it with an indefinite and even infinite universe bound together by fundamental components and laws. All these components are therefore, he said, placed on the same level of being. This in turn implies the discarding by scientific thought of all considerations based on value concepts such as perfection, harmony and meaning: and finally, the utter devalorization of being; the divorce of the world of value from the world of facts (Koyré 1957: 2).

However, this devalorization of being did not include and has not included human beings. Our 'revalorization' of ourselves has been based precisely on our assumption of an Archimedean view of the earth; on our being able to take a 'cosmic, universal standpoint' from which to assess our role and place in the universe. As we have seen, that role has gradually become that of potential destroyer of all life on the planet – or creator of new life. And more alarmingly still, that destruction now needs no other instrument nor invention other than our continuing to pursue a high energy, carbon-rich, resource-wasting lifestyle.

In a major book entitled *The Anthropic Cosmological Principle,* distinguished astronomer John Barrow and similarly distinguished professor of mathematics and physics Frank Tipler offered a collection of ideas as 'a means of relating Mind [*human mind*] and observership directly to the phenomena traditionally within the compass of physical science'.

The word *anthropic* here has the same function as *homo,* that is, one may assume that it is intended to refer to the whole human race. So the title suggests that the existence of our species constitutes a principle in relation to the existence of the entire universe. Therefore the principle itself connects 'Mind' (human) and our capacity to observe and understand what happens in the world with our unique status (observership). This principle is then taken as fundamental to the evolution of the universe. In other words, the goal and purpose of the universe is taken to be our rational, scientific capacity to observe and compute that evolution. The destiny of the universe coincides with ours: which is to have 'minds' uniquely capable of observing it.

The term 'observer' denotes an attitude of scientific non-attachment to what is being observed, so there is an implicit assumption that our 'observership' extends to and operates anywhere within the universe. Geocentrism no longer limits our horizons. In Arendt's words, 'man cannot possibly be of this world even though he spends his life here'. True, astronautical travel appears to support this assumption, but a moment's reflection reminds us that space travellers take with them the earthly atmosphere and physical resources their bodies need. They cannot (yet?) live outside their terrestrial environments.

Barrow and Tipler said that they regarded as axiomatic the Copernican principle which holds that we do not occupy a privileged position in the universe. However, they also said that 'like most generalizations it must be used with care' (Barrow and Tipler 1986: 1f.). They go on to single out 'Mind and observership' as making the significant difference in respect to our position; and cite our ability to compute the size of the universe as displaying an intrinsic bias toward our evolution as 'minds' capable of observing and computing it. This argument for our significant difference takes exactly the same form as that used by theologians, although they cite our possession of 'a soul' as making us special.

The latter do not, however, as far as I know, go on to cite the soul's computing abilities as proof of their arguments. Instead they base their claim on the soul being like to God and therefore unlike any other creature on earth. Barrow and Tipler, however, appeal to no referent outside ourselves. This makes their argument circular: for they hypothesize 'Mind' of a particular sort as special to us and then use this 'Mind' to make us special within the universe, because we have worked out satisfactory hypotheses about the universe – satisfactory, that is, to our minds. What we do with these hypotheses in regard to the quality of life on earth for all living beings remains unremarked, let alone acknowledged.

The Gaia principle

But our experience of the input from the Hubble Space Telescope is now demonstrating how inadequate those homocentric hypotheses are. No privileging of scientific epistemological processes can establish a claim to human omniscience in regard to the size or nature of the universe. My concern, however, is not with the semantics of the anthropic principle but with the fact that it makes observership, a particular form of homocentrism, into a principle of purpose in the universe. By doing this, one intellectual form of anthropocentrism is being used to discount another principle of purpose of critical importance, based on our geocentrism. That is that for good or ill, our purposes and activities, especially over the past two hundred years or so, aimed at increasing monetary wealth through sustained economic growth and coupled with a human population increase unmatched in our history, are central to radical changes in global climate that affect and threaten the fabric and equilibrium of life on earth.

Furthermore, we ourselves are now seen to be so endangered by these changes that, unless we come down from Archimedean observation of the earth and start to pay much closer attention to what is happening there, our own survival in the universe becomes increasingly unlikely. This situation calls for us diverting our energies, imagination and resources into playing a central role in dealing with the causes and effects of climate change,

These truths have emerged from ecological, earth-based life sciences that use rigorous observation techniques to analyse those causes and effects. These scientists too have come to see how central a role we play in them. A while ago I heard an ecologist remark, with a mixture of astonishment and gratification, that '*people* are now included in ecosystem evaluations!' However, that inclusion has yet to penetrate a public consciousness conditioned by urban living. Over half the projected world population of seven billion by the end of this century will live in cities. Our essential inclusion in, and effect on, natural life support systems remains at odds with what we observe and with how we maintain our lifestyles within cities. These subvert any view of ourselves as being dependent on the natural world. Rather, the prevailing presumption is that we are uniquely entitled to manage the world for our ends alone or, indeed, leave its confines and, with Hubble, 'fly off into the hugeness of the universe'. This Archimedean viewpoint is all too easily taken for granted and/or adopted in the skyscrapers that now characterize the skyline of global capitalist institutions and the mindset of decision-makers within them.

The reality is, however, that like all terrestrial species, we are products of, participants in and dependent on the life support systems that have evolved throughout earth's history and that make earth, and earth alone, habitable for us. This truth and its challenge to our presumed observer status, whether on earth or outside it, is implicit in James Lovelock's Gaia theory, based as it is on the ecological principle that the ability of any terrestrial organism (including ourselves) to emerge, increase or maintain its numbers and to spread geographically (not universally) is limited by, and dependent on, environmental factors, particularly the availability of natural resources and energy capacity. This reality raises the fear that we may be unable to deal with these limitations and their increased demands on us in a sufficiently speedy, creative and just manner. The hope must be that we can and do.

We can still feel, says Koyré, the excitement and pride glowing beneath the cool and sober wording of Galileo's *The Message of the Stars*.

> In this little treatise I am presenting to all students of nature great things to observe and to consider. Great as much because of their intrinsic excellence as of their absolute novelty, and also on account of the instrument by the aid of which they have made themselves accessible to our senses.
>
> (Koyré 1957: 89)

That excitement and pride is felt today by those scientists who 'observe and consider' all the 'great and excellent things' to be seen at microcosmic level, on earth and in the oceans, through an electron microscope. This has, says Lynn Margulis, shown the folly of considering people as special, apart and supreme.

> The power of consciousness, of our society and our technical inventions has made us think we are the most advanced form of life on the planet. Life on Earth has traditionally been studied as a prologue to humans ... But during the past three decades, a revolution has taken place in the life sciences. The microscope has gradually exposed the vastness of the microcosm and is now giving us a startling view of our true place in nature ... Far from leaving micro-organisms behind on an evolutionary 'ladder', we are both surrounded by them and composed of them ... This realization sharply shows up the conceit and presumption of attempting to measure evolution by a linear progression from the simple – so called lower – to the more complex (with humans as the absolute 'highest' forms at the top of the hierarchy).
>
> (Margulis 1986: 27f.)

The miracle that saves the world, the realm of human affairs, from its normal, natural ruin is ultimately, says Arendt, the fact of natality. That means, in this instance, the birth of new women and men and the action they are capable of by virtue of being born (Arendt 1958: 247).

6   The Givenness of Events

The ordering of the universe;
the operation of the elements;
the beginning and end and middle of times.

(Wisdom of Solomon 7: 17–18)

We are now living in what for our species could be seen as 'the middle of times': that is, a long time after the events denoted as 'the beginning' when the order of the universe and the operation of the elements were established. Compared to our distance from that event, a relatively brief period of time has passed since the historic ones discussed in previous chapters. The givenness and effects of all these events taken together are part of the givenness of our lives now, with 'the operation of the elements', the dynamic changes in global climate, taking on an unprecedented importance. Already these changes are affecting some of us profoundly, especially as they are closely bound up with our past and present activities. Indeed we are told that the transformations we have wrought in their operation may precipitate 'the end of times' for our species. The givenness of human-induced changes to the climate and their effects has become a present concern as well as part of the givenness of future lives.

In the verses from the Book of the Wisdom of Solomon quoted above and those following them, while Wisdom is part of the divine order, and divine in herself, she is also part of the human order. She is presented as replete with the knowledge of nature and all its workings and able to communicate this to us. In a contemporary idiom, the author would praise her knowledge of earth sciences, meteorology, astronomy, zoology, psychology, botany and pharmacy.

> She is concerned with the world as humans live in it and understand it. She is teaching them about the nature of the universe ... All peoples and all beings experience alternations of solstices, changes of season, and so on. Here is a view of Wisdom that integrates the world.

(Long 1992: 40–3)

Reciting and learning to understand the sequence of primary events is a call for us also to be wise in regard to them. This means that whatever knowledge we have of them must be used to find the best ways in which to correlate our lives with them for the benefit of all in both our own and future generations. Our knowledge and understanding of the events at the beginning of time has necessarily been transformed over the years. Now its extraordinary, unconditional, originary nature, and particularly our understanding of the operation of its elements, emerges in a contemporary idiom in James Lovelock's description of the genesis of Gaia:

> In the beginning
> The Earth evolved chemically and physically.
> Sometime after its birth
> The first living organisms appeared,
> Probably at a single place.
> Gradually
> Life spread over most of the planet.
> It was mostly ocean.
> During this period life and Earth evolved separately.
> As life grew abundant
> It began to change the environment
> Until its evolution and the Earth's evolution
> Merged into a single process:
> The dynamic system
> Gaia.
>
> (Personal communication, April 1999)

An important feature of Lovelock's account of the emergence of Gaia is that – as in the biblical Genesis text – the potential for the givenness of life lay first in 'the waters'. Scientists generally agree that it was from this 'deep time' ocean – still covering seven-tenths of the planet – that living entities did in fact emerge. Their emergence was made possible through the presence in the waters of life-giving properties that had evolved through change and interaction powered by some form of death as well as life.

This constitutive interaction between life and death was traced back to its source in the atoms that compose water by scientist Tyler Volk. The hydrogen in it was made about thirteen billion years ago. Its other component, oxygen, was forged in a series of fusions that took hydrogen into heavier elements. Here, he says, is a case in the realm of physics of 'death, thus life'. For in a sense, the hydrogen dies to form oxygen.

> This life from death is life of another entity, oxygen in this case, that later combines with some of that abundant, still living, primordial hydrogen, *giving birth to water* that now nourishes algae and my life. In looking at how

death becomes life we invariably have to shift scales in types of entities, often upward to the larger encompassing context.

(Volk 2002: 224f.) [Italics mine]

Characteristics of the givenness of Gaia

The givenness of the larger encompassing context of Gaian processes, through which death becomes life and in which death nourishes life, is discerned by us in different types of entities and at shifting scales. It is present in the multilevel modes of life brought forth and nurtured on earth. This givenness coincides with the novelty, the beginning, the 'something new under the sun' that is the genesis of all life. It is the mysterious creative operation of elements that has made every gift event and will make every future gift event possible. Its absolute precedence, its presence as the 'ever-enduring base of all things' signals its anteriority to all present life forms and its irreducibility to any one element or mode of living.

It also signals our intrinsic relationship with, dependence on and sensitivity to the original elements and conditions – such as the sun's energy and the composition of the earth's atmosphere, oceans and soil – that were the essential 'given' for our emergence and our continuance as a species: just as they remain the given for all emergent life forms on earth today. That dependence and sensitivity also marks the interactive, correlative and shared character of our lives. The gift of life has never stood alone in some relation-free zone. It always signifies some relationship, some connectedness and interaction between receivers and givers. No being, at any moment in its life, lives only to, for or by itself. No one dies to, for or with herself alone. This symbiotic givenness intrinsic to life means that each being is generated with, chemically modulated through and sensitive to other modes of being, whatever, whoever and wherever they may be (Primavesi 2000: 2–9).

It also means that although we were not there when the shell of the mystery of life first broke upon the earth, we have received and continue to receive its gifts, transformed by the lives and deaths of those who have lived before us. We know that the fossil fuels that now play such a crucial role in our lives come from sedimentary deposits formed by the remains of billions of former living beings. These have shaped our landscapes and now play a part in shaping our lives through their extraction and use, whether in coal or oil as fuel, or in salt as a life-sustaining mineral deposit. We have inherited these lives in a form that has metamorphosed over billions of years. Similarly, in the course of human history, we inherit the physical, cultural, literary and genetic remains of our ancestors. And now we are inheriting their input into, and changing of, the global climate, just as future generations will inherit and be influenced by our inputs into it.

The givenness of climate events

The four nominally distinct yet interconnected events of the preceding chapters were shown to have contributed to the givenness of our lives now. Together with all other events before and during human history, they constitute a givenness described by Deleuze as:

> [A]n impersonal and pre-individual transcendental field – a nomadic distribution of emissions of singularity radically distinct from fixed and sedentary conditions.
>
> (Deleuze 1990: 118)

This is clearly true of climatic events and their effects. The words 'impersonal', 'pre-individual' and 'nomadic' – with the term 'emissions' now ringing rather differently in our ears – accurately describe the dynamic global processes through and from which those historic events emerged – as will all such events in future. We know that our personal, individual 'emissions' and those generated by events in which we participate will affect future earthly conditions and the living beings and events situated within them. What happens now, in one place at one time, intersects, interacts and interweaves, sooner or later, with weather events elsewhere.

It is in this context that the ambiguity of gift events becomes apparent. For in the relationship between present givers and future receivers, the effect on the receivers is, in some measure, proportionate to the donors' goodwill and concern. Yet the fact that we shall never meet those receivers nor be affected by nor be able to predict precisely the climatic conditions resulting from the present contributions of our present lifestyle to them makes it difficult for us to take responsibility for those putative effects as seriously as we ought. So while we are being constantly reminded about the impact of increased carbon emissions on climate in the near future, the evidence of our senses does not reinforce the message.

This problem relates directly to the givenness of climate and our own relationship with it. We experience and view it now 'after the event' of its formation. For us, it's a done deal. Its 'givenness' ensures that our personal detection and attribution of its distant or even proximate causes always works retrospectively: its literal 'unpredictability' remains experientially true.

It is true scientifically too, albeit to a lesser extent, since the data we work with is necessarily gathered prior to the weather forecast. Climatologists have constructed a 'Virtual Earth' computer model that brings together a vast array of understandings of how the earth's atmosphere, oceans and life actually work, integrating these findings into models that provide accurate weather forecasts for particular areas on a daily basis. 'Virtual Earth' can also be subjected to changes in the concentration of greenhouse gases, in forest cover and in the amount of energy received from the sun, establishing the effects of these acting together and in isolation from each other. Presently, they project further warming of between 1°C and 6.4°C by the end of this century and sea level rises at between 28cm and 58cm (Betts 2007: 49–51).

Another problem hampering responsible reaction to climate change is that we perceive and experience it only in part: from a particular perspective conditioned by time and place. Its overall givenness is simply beyond our grasp. Yet we know that while we experience and analyse particular present climatic events or predict future ones, their causes and effects are globally integrated, as they will be for those who come after us. This means, therefore, that our local sensitivity and response to the given conditions of today's weather becomes part of the givenness of the global climate anterior to all lives in the future. It's cold today, so we turn up the thermostat. It was too hot yesterday, so we turned on the air conditioning. These individual actions have 'impersonal' consequences that we can neither assess nor predict.

So Mark Twain's quip that 'everyone talks about the weather but no one does anything about it' will certainly apply to us retrospectively – if we continue to do nothing but talk about it. As a species that has grown to proportions unprecedented in earth's history, our responses to it have become powerful enough to affect the 'givenness' of our own and others' lives – for better or worse. On a planetary scale we have become a geological agent actively reconfiguring earth's crust and climate: one rivalling volcanoes, glaciers, wind erosion and water. This reconfiguration has happened and is happening through industrialization, mining, damming of rivers and accelerated soil erosion – all accompanied by massive population growth (McNeill 2000: 21–39).

Our most direct 'give-aways' to earth's crust and climate, in the form of consumerist waste, carbon emissions and greenhouse gases, now appear increasingly poisonous. And as the cumulative effects of our previous overuse of energy resources become perceptible in weather variations above the norm, and are presently seen in such events as the shrinking of the polar ice cap, we can no longer shirk our responsibility for the future effects of present overuse. At the same time, growing awareness of the fact that those present effects are rebounding *on us* demonstrates not only that we participate in climate regulation but also that, as just one species, albeit a relatively powerful one, they are neither determined nor constrained by our needs or desires alone. We do not own the earth's elements that create the global climate nor do we determine their operation and outcome. We participate in weather production – and not on our own terms.

In this situation, our input into the givenness of global climate events evokes the following questions:

> What does giving mean? What is at play in the fact that all is given and how are we to think that all that is, is, only in as much as it is given?'
>
> (Marion 1998: 38–9)

We can start by accepting that, as part of 'all that is' at this stage in history, our bodies are not defined by the form we see in a mirror, nor by our name, voice, or actions.

They are *defined only by a longitude and a latitude*. In other words, by the sum total of material elements belonging to them under given relations of movement, rest, speed and slowness (longitude); the sum total of the intensive affects they are capable of at a given power or degree of potential (latitude).

(Deleuze and Guattari 1988: 260)

But we must go on from that rather narrow but essential definition to look at how we act under those material conditions: at the modes in which we live out our relationship with all the earth's inhabitants and with its material elements. In other words, at the quality of interactive relationships between us as givers and receivers, between givers and gifts and between gifts and receivers in gift events: at how we react to our dependence on others and use the inheritance from those who have gone before us.

But an inheritance, by its very nature, can only be passed on and acquired by others through death. That is as true of our collective bequests to those coming after us as it is for individuals. Our death is an essential part of the gift events that will be received by future generations.

The givenness of death

Death is a given for all of us, wherever we find ourselves in life. The origins of death do not lie in any one event in human history, although our violent death-dealing reactions to each other, as in our colonial history and its aftermath, can be seen as the ultimate rejection of it as a gift. Yet death plays a natural, essential role in the givenness of all life on earth and contributes to the originary givenness of all our lives now. As we are unable to photosynthesize the sun's energy ourselves, we depend bodily on the cessation of visible life in the plant and animal world where that process goes on. That is part of the givenness of being human.

It makes us, at best, sensitive to modes of death and how they affect other living beings. It makes us sensitive to the fact that the primary ambiguity of the gift of life-death is inextricably linked to the givenness of every event in our lives. 'Unless the seed dies, it remains alone.' Yet in traditional Christian teaching God's gifts have become almost entirely identified with bestowing life. Their ambiguity was preserved in different Germanic languages where the German word '*Gift*' once meant both 'present' and 'poison'. For the modes in which death naturally occurs are always bound ineluctably to those of life. You have first to be alive and experience life in order to die – or to inflict death on others.

This makes death both mysterious and feared because all too often we experience it as tragic and painful. But that does not mean that it is not a gift given by Gaia (or by God, ultimately, however we understand God). Nor may we imagine that death is a punishment imposed on us by God and not (as in the generation of oxygen) a gift integral to that of life. The arrogance, indeed *hubris* implicit in the Christian claim that death stems from the act of a single human being, Adam, requires the belief that the single act of a single human being could, and *did*, change the nature of the givenness of the seminal event of creation: of the very structure and operation of the universe itself (Pagels 1990: 128–33).

The same arrogance is evident in the scientifically based anthropic principle and its claim to our uniqueness and immortality in the universe on the basis of intelligence:

> [I]ntelligent information-processing (that is, human beings) *must* come into existence in the Universe, and once it comes into existence, it will *never die* out.
>
> (Barrow and Tipler 1986: 1–16, 21–3) [Italics mine]

Such arrogance obscures and indeed denies the more inescapable scientific fact that our dependence on the original conditions of the creative structure of our world, on the gifts of life and death and on the interactive processes and relationships between life forms that sustain our living, calls for an appropriate and considered response from us. At the same time it limits the nature of what we are and can do. Gaia's overarching role in begetting and nurturing life, accompanied by death as its natural conclusion, is acted out personally by us in the variety of roles we play out in our communities. Inflicting violent death, in various modes that can be, and are, registered along a scale of their ability to cause pain and suffering, now marks us out not as the most intelligent species on the planet but as the most lethal.

Transformative roles in life

Our contributions to the nurturing and well-being of other lives, by virtue of our shared grounding in Gaia, are personified for me in the Gaelic figure of the mother-goddess of landscape known as the *Cailleach Bhéara* or Supernatural Female Elder. Her traditions are found attached to natural features of the physical landscape – mountains, lakes, rivers, tumuli, caves – whose shape she has moulded and whose location she has fixed. At the learned, literary level, she is seen as the personification of territorial sovereignty. But it is in her presence in popular tradition that her autonomous creative potential resides:

> In her person she constitutes an overarching female matrix of sovereignty and fertile power that is as vast and untameable as the wild, wide landscape and is yet as nourishing and as intimately fruitful for human beings and for human existence as are the services of the *bean ghlúine* (midwife), the *bean feasa* (wise-woman) and the *bean chaointe* (keening-mourning woman), to name three human female personae whose *cailleach*-inspired (and derived) performance of service to the community was so essential.
>
> (O Crualaoich 2003: 28–9)

The roles here discerned as essential for the well-being of a community are personified as those that bring new life into the world, that teach the wisdom required to nurture it and enable others to live well and happily and then to accompany others in death: mourning them by remembering and carrying on what they accomplished in life. All these roles point to key moments in our lives – to their beginning, middle and end – when the services of others are essential in building up not only our lives but also the lives of those around us and those who will come after us.

In the context of climate change, affirming and enhancing the beneficial potential of our presence means being able, willing and competent to act wisely. The role of the *bean feasa* is that of the biblical Wisdom figure, concerned with the way humans live in and understand the world. She carries out her role by providing a model for living wisely, in the light of our understanding of the nature of the universe and our belonging within it.

> For in her there is a spirit that is intelligent, holy, unique, manifold, subtle, mobile, clear, unpolluted, distinct, invulnerable, loving the good. Beneficent, humane, steadfast, sure, free from anxiety.
>
> (Wisdom of Solomon 7: 22b)

Asphodel Long gives a detailed account of what each of these attributes involves and calls for in us (Long 1992: 46–55). While this model of Wisdom, like all models, epitomizes absolute qualities, those attributed to her make a present and future contribution to the well-being of the community: a contribution now more vital than ever. By virtue of our belonging within Gaia, those contributions are effectively transpersonal, transnational, indeed, transmundane. Therefore, when we are wise in conducting our relationships with those near to us, it contributes toward environmentally healthy relationships within the whole.

One of the increasing effects of a raised awareness of climate change is that we are becoming wise not only to our dependence on and shaping by past weather events but also to the effects of historic events on the shaping of relationships between us. With this in mind I shall take a closer look at the ways in which events associated with colonization have influenced them. Their historic legacy has undoubtedly reinforced structures of alienation between us that in turn have legitimated our inflicting violence, war and death on others and degrading their environments. Continuing with that pattern of behaviour rather than consciously alleviating its effects plays much too large a part in the givenness we are handing on to future inhabitants of the earth.

The legacy of colonization

A previously unmentioned aspect of this historic legacy is the type of justification offered by colonizers for inflicting violence and death on the colonized. Often this is made by using stereotypes that identify a particular class of person as 'needing' the 'benefits' of colonization. Once again John Locke offers a pertinent example in his *Two Treatises on Government*.

Those who live under the law are civilians: those who live beyond it are barbarians. 'Law makes men free in the political arena, just as reason makes men
free in the universe as a whole.'

(Deane 1983: 5)

Seamus Deane uses this statement to open his discussion of the colonization of
Ireland. He shows that the stereotypes 'civilians and barbarians' were already well
established there when Locke wrote – with the Irish playing the role of 'barbarian'. The implied justification behind the stereotypes was that the strife in Ireland
(that is, a series of Irish rebellions against English rule) was the consequence of a
battle between English civilization, based on laws, and Irish barbarism based on
local kinship loyalties and sentiments. The added complication of religion
(Catholic) was considered as helping intensify Irish barbarism by fostering ignorance and sloth, disrespect for English law and respect for Papal decrees instead.
All of which would add up to 'irrationality' as opposed to the reasonableness of
law-abiding English civilians.

Deane comments dryly on the thinking behind this categorization:

The wisdom of English commentators on Irish affairs has always been viti-
ated by the assumption that there is some undeniable relationship between
civilisation, the Common Law and Protestantism.

(Deane 1983: 7)

Living in England twenty-five years after Deane wrote, when Irish relationships
with England are more peaceful than at any time in the history of either country, I
am particularly struck by the disturbing analogy he draws between the stereotypical
barbarian seen from the viewpoint of an English civilian and a potentially criminal
type: above all, the *politically* criminal type that is your neighbourhood terrorist. All
too often here, after the London Underground bombings, he is seen to have all the
classic faults of the barbarian as seen by civilians. But he is Arab not Irish, Muslim
not Catholic, a member of an extreme Muslim rather than a Catholic sect:

He draws money from the benevolent state which he intends to subvert ... is
from an area of dirt and desolation not to be equalled in western Europe.
Finally, and worst of all, he is sometimes a she.

(Deane 1983: 12)

This startling analogy tells us something about the postcolonial legacy of today and
about 'civilian' reactions to that legacy. Kwok Pui-Lan highlights the way in which
the 'discovery' of America changed and, at the same time, exposed the world view
of Europeans. Christians encountered peoples whose cultures and ways of life dif-
fered radically from their own. Up to then they believed that human beings were all
descended from the same family tree. Now they had to reckon with, and account
for, the differences between themselves and those others. In many cases their first
reaction was to doubt that these indigenous peoples were human beings.

If they were, she said, they were regarded as imperfect or incomplete ones – a statement that correlates perfectly with Deane's analysis of barbarism in that this attitude predates the colonization of America by the British. In that case the indigenous peoples were seen as infidels to be cajoled or forced into the Christian community. Their nakedness symbolized their alterity, their alien and alienating status: 'that of the "Other" at the periphery of the "civilised" European world.' From the beginning, their otherness was replete with sexual overtones, establishing a relationship between Christian imperialism, colonialism and sexual violence. They were often likened to the Canaanites in the Bible, in the sense that they were seen as descendants of unsavoury sexual relationships and the personification of sexual perversion (Kwok 2005: 14–15).

With Deane, she recognizes the importance of the acts of identity formation at stake in both the colonizer and the colonized, particularly those employed by Christian colonizers. She highlights a paradigmatic exchange between Regina Schwartz and a student reading the Bible. 'What about the Canaanites?' Yes, what about them – and the Amorites, Moabites and Hittites? What about all the peoples and their gods other than the Israelites and their God?

> Through the dissemination of the Bible in Western culture, its narratives have become the foundation of a prevailing understanding of ethnic, religious and national identity as defined negatively, over against others.
>
> (Schwartz 1997: ix–x)

Regina Schwartz notes that the themes of 'covenant', 'chosen people' and 'promised land' have been used repeatedly by the Christian West to justify colonization of non-Christians and annihilation of indigenous peoples, accompanied (of course) by the appropriation of their lands. Indeed, part of the givenness of Western Christian history is what she calls 'the apparently compelling myth of identity: that the divine promise of land to a people creates them as a people'. The story of Israel proper begins, she says, with Genesis 12 when the people of Israel are formed and promised numerous descendants, a mighty nation and land. But this gift is subject to a contract: loyalty to God. Faith in the deity is the guarantee of a land grant. A people are a people united by collective memory of a promise that they will possess a land.

> Promise is the key word here, for that possession is elusive ... As if anticipating the perils of the idea of attaching land to identity, the biblical writers have also included a critique of it ... A closer look at that intimate etymological relationship between man, *Adam,* and land, *adama,* reveals that it is between human beings and all land ... This use of *adama* marks a departure from the dominant idea that a people are specified by a particular land, for it suggests that if man is a creature of land, he is an 'earth-creature' who is not tied to one piece of it.
>
> (Schwartz 1997: 42–4)

There is, she says, a dangerous consequence to attaching identity to territory: when a people imagines itself as the people of a given land, the obvious threat to that identity is loss of that land. Precisely that fear drives the plot of biblical narrative. Indeed she suggests that it is likely that only when the Israelites are landless do they imagine themselves as a people who have inherited a land. Built into the logic that imagines Israel as a landless entity is also an Israel that *is* a landless entity. (This fits perfectly into the mindset of the 'Pilgrim Fathers'.) Deep inside that logic may be the admission that possession of the land is an impossible fiction, if by it we mean having exclusive rights to it (Schwartz 1997: 44).

Her insights about 'possessing' land look back to what was said in Chapter 4 about its *legal* appropriation as 'property' and then its metamorphosis into money. It also looks forward to what will be said about the nature of giving in the following chapters and, in particular, to the image of God underlying the capitalist logic of property possession. Here I want to highlight the continuing effects of the legacy of colonization for both colonizers and colonized, with some of us playing both roles at different times. Our memories of it, from either perspective, play a part in constructing the identities of both by each other.

> The biblical preoccupation with memory (the various declensions of *zakhar* appear 169 times in the Hebrew Bible), the proliferation of narratives that record memories and the numerous explicit injunctions to remember are all in the service of forging a community through the creation of selective memory.
> (Schwartz 1997: 143)

As we have seen in regard to reactions to terrorists in Britain today, wherever we live, the givenness of our lives and of our identity has been transformed and is coloured by memories of colonialism. Its present manifestation, in our identification as consumers, creditors or debtors within a world marketplace, is the creation of a very selective memory about the right to exploit the riches of the earth for individual profit. That particular historical phenomenon has left us addicted to violence and our landscapes marked irrevocably by it. It surfaces openly in the social, political, economic and environmental fallout of the Cold War and in the battles fought by its surrogates in the postcolonial lands of the Middle East – most notably for the possession of energy resources in oil and natural gas.

Colonialism's legacy to women

One aspect of that legacy deserving special attention is the violence done to women that is implicit in the process of colonization and explicit in the particular nature and effects of their oppression. Deane's comments on the classic faults of the stereotypical barbarian, as seen from the civilian viewpoint, have as rider the throwaway, but none the less significant remark: 'Finally, and worst of all, he is sometimes a she' (Deane 1983: 12).

Kwok Pui-Lan's detailed study of particular effects of postcolonialism on women gives a comprehensive account of what that 'worst of all' actually means from a colonized woman's perspective. She rightly points out that many white feminist theorists and theologians give priority to gender oppression over other forms of discrimination such as racism, classism, heterosexism and colonialism. Patriarchy is seen as the root cause of societal problems and white women as victims rather than oppressors. She sees this as a complicit attitude: one that does not take into consideration the interlocking nature of oppression and so obscures how power actually operates in society.

> If we look back to some forty years of history of Euro-American feminist theology, we cannot say that anti-colonialism and anti-imperialism have been its major concerns. While people outside America have protested American imperialism for decades, those inside, feminists included, have been slow in debunking American empire building except during the last several years with the government's declaration of a 'war against terrorism' and the invasion of Iraq.
>
> (Kwok 2005: 130)

Perhaps because British colonialism in Ireland is part of my historic legacy, reading Deane not only clarified for me the role played in colonization by identity formation in both colonizer and colonized, but my involvement with feminism also alerted me to his afterthought that the 'faults' of the colonized or terrorist are worsened by being female. Who regards this as a 'fault'? The colonizer? If so, how are women supposed to deal with it?

Kwok's point is that white women theorists (that is, those descended from colonizing races and those assimilated into them) have not dealt with it. Their efforts have been directed at and from within societies in which women are legally subordinate to men; at assumptions that they are physically, psychically, morally and intellectually inferior to them. The extensive references in Kwok's book provide (for those who need it) an introductory course on the origins, extent and effects of this effort and her pertinent quotations serve as a refresher course in diagnosis of its shortcomings. As a Christian theologian she highlights the different ways in which it has influenced, indeed circumscribed the thinking of feminist theologians who, having criticized masculine images of God, project an image of God as Mother that merely reinscribes sexual difference because only her maternal and nurturing roles are stressed (Kwok 2005: 130–6).

The specific role of colonized women, however, continues to be that of the slave or bonded labourer bent low over her work and, all too often, carrying a child: one who will in turn be bound to this role. This is a far cry from that of the contented, nurturing mother invoked in Western society. Nor does it correspond to those associated with *Cailleach Bhéara* who is midwife to life in all its forms: personifying the hope implicit in natality. She teaches us how to support life through living wisely and helps us cope with the sorrows of death. Together with her role in directing and teaching the members of the community how to live well and happily, she symbolizes the integration of life, living and death and the receiving and giving of different forms of vitality that bind us together through life. This figure forms part of my cultural inheritance and has, in some fashion, survived the colonizing process.

However, as Kwok points out, the symbolic order of any given society may or may not be symmetrical with the social order. So when the female symbolism of nature is restricted to women's reproductive roles in fertility and nurturing, all too often that reinforces the domesticity of women. Then the symbolization of the feminine in connection with nature serves to justify women's subordination. The ideology of colonization included the symbolization of foreign land as a female body to be possessed and conquered. Kwok quotes Edward Said:

> Such imagery is connected to the configurations of sexual, racial and political asymmetry underlying mainstream modern western culture.
>
> (Kwok 2005: 226)

That asymmetry bears down particularly hard on women in former European colonies. Before and since the United Nations Conference on Environment and Development in Rio in 1992, socio-economic analyses, feeding into and emerging from the Conference, highlighted four classes of people most at risk from environmental degradation – and now from the effects of climate change. They are the poor, indigenous peoples, women and children. These four classes were also seen to be, and still are, the most powerless to do anything about this degradation. It is also significant that women in fact belong to three out of the four groups.

Women's reaction to climate change

In the policy documents issuing from the Rio Conference the most pertinent to this state of affairs was Chapter 24 of Agenda 21. Its utopian objectives included the participation of women in national ecosystem management and an increase in the proportion of women technical advisers and planners. Utopian they may have been, but they did recognize the multidimensional character of oppression and how different forms of it are related to one another.

Fifteen years later, at the 2007 UN Climate Conference in Bali, a global alliance of women for climate change justice gave a position paper on 'Gender Issues in a Carbon Calamity World' that made the same point (http://www.gendercc.net). It pointed out that not only the impacts of climate change but also certain activities proposed to mitigate its impact can be a calamity for women. Harking back to issues raised here in Chapter 4, the presenters drew attention to the effects on women of market-based solutions, such as the Clean Development Mechanism, that allows public and private actors from industrialized countries to generate emission credits through climate protection projects in developing countries.

They pointed out that the benefits to women are very limited, since the bulk of investments go to large-scale power generation or industrial projects rather than into energy efficiency in the domestic sector or into small-scale renewable energy projects for rural communities. A number of projects, such as large-scale monoculture plantations for biofuels, are even harmful to those communities, for they require destruction of forests in tropical countries. Landfill gas utilization projects lead to expansion of landfill sites in some of the poorest areas rather than their closure.

The women meeting in Bali demanded that the real, direct and underlying causes of deforestation, such as over-consumption, agro-fuel expansion, fossil fuel extraction and the lack of respect for indigenous peoples' rights, be addressed. They also vehemently opposed the inclusion of nuclear energy projects in the Clean Development Mechanism in order to make them eligible to receive financial incentives, pointing out that climate change should not be combatted with hazardous technologies involving uncontrollable risks for future generations:

> We need to question the dominant perspective focussing mainly on technologies and markets, and put caring and justice in the centre of the measures and mechanisms.
>
> (http://www.gendercc.net)

If, then, as the women ask, we must look beyond market-based solutions and a market-based consumerist economy to deal with the problems of climate change, what kind of economy are we talking about? The following chapters will outline a theology based on an economy of gift events.

7 The Economy of Gift Events

God has given the world to man not only as a gift of continuous fruitfulness, but as one immensely rich in possible alterations, actualized by each person through freedom and labour. This actualization, like the multiplication of talents, is the gift of humankind to God.

(Miller 2000: 62)

This quotation from Orthodox theologian Dimitri Stăniloae provides a different perspective on the world from that of a capitalist economy by envisioning our lives as gift events. They are characterized by actual relationships of receiving and giving dependent on the givenness of the original gift of the world by God. That seminal gift event was pre-original to our emergence and its givenness made our emergence possible. For it contains what Stăniloae calls the 'undefined potential' that, through our freedom, love and labour, we and all living beings transform into gifts to God. We, however, can and are encouraged to do this consciously, seeing this divine-human interaction in and through the stuff of which the world and we ourselves are made. This is the perspective from which a theology of gift events based on a divine economy of love can be envisioned.

His use of the word 'labour' in regard to our transformative activities may appear reminiscent of that understood and proclaimed by the Protestant reformers and by Locke. But in contrast to that Christian tradition which sees human labour as a consequence of our estrangement from God and therefore as a negative factor in human experience, Stăniloae argues that work is spiritually valuable, positive, even joyful. For thought, imagination and physical labour all contribute to the transformation of the natural world into human gifts to God. This is an essential part of the dialogue of the gift that lies at the heart of our proper relationship to the natural order.

No one returns to God the things he has received without his own labour being added to them.

(Miller 2000: 63)

But our gifts to God must be freely given. The quality of freedom in what we do is decisive for this theology. For while God gives us the ability to work, it is within our power not to use that ability. We are free to choose whether or not to increase the talents given us.

While this perspective on freedom and work may appear religiously utopian, it is worth noting that it emerged out of the political circumstances of Stăniloae's life. It was an attempt on his part to meet in a positive way the challenges of a Marxist ideology in which a human being, the *homo sovieticus* of that ideology, was regarded primarily as 'the worker', whose task was to transform and humanize nature by his labour. Today the task of 'the worker', the *homo capitalisticus*, is to transform nature into monetary profit that can be used to drive the engines of consumption and waste of the resources of the natural world. So while the trajectory of Stăniloae's theology of the gift is transcendent and Godward, it also had, and still has, a very important horizontal, human dimension. For within the present context of a highly industrialized and technologized capitalist society, our labour is consistently downgraded, ignored or exploited for corporate monetary profit (Miller 2000: 63–4).

The mystery of giving

Stăniloae's Godward trajectory, together with its horizontal, natural dimension, is analogous to the ecological patterns and processes described in this book that systematically connect human participation with the givenness of Gaia. At whatever level of understanding we approach this givenness, however, any expression of our insights into, or record of, this ultimately unfathomable region is inadequate. Poet Brendan Kennelly rightly describes it as 'the mystery of giving': one that always resists what he calls 'the feared definition' (Kennelly 2004: 23–4).

Therefore, defining the roles played by giver, gift and receiver in these processes would be (to use Flannery O'Connor's image) like trying to describe the expression on someone's face by saying where the eyes, nose and mouth are. Pertinent too is Martin Buber's comment that a melody is not composed of tones nor a statue of lines. What can be done, however, is to describe the relationships between those playing these integral roles at particular moments. Always, however, with the proviso that the import of the relationships themselves far exceeds any description of a particular moment between those involved. Their location within earth, the original ark, makes it the shared space where the relational process of receiving and giving occurs, recurs and continues.

Nevertheless, naming this all-encompassing process as the mystery of 'giving' (rather than receiving) emphasizes the role of the giver. The reason is that, from our perspective, there was an initiating moment in the process that we see now as a free act of giving. At its most personal and obvious, it was the moment when each of us was given life and birth. But that moment belongs within, and depends upon, the undefined potential of a continuum of gift relationships whose earthly beginnings are shrouded in the mystery of the seminal event of creation. One such initiating moment was when the first photosynthesizers caused oxygen to become a constituent of earth's atmosphere. Without it the earth would now be as dry a desert as Venus or Mars. Its presence in the air prevented the hydrogen of water from escaping into space, as it does on those planets (Lovelock 2007: 1).

The fact that this act of giving took place at a moment in geohistory that is beyond our ability to pinpoint both underlines and underlies the mystery. Where, when or why oxygen became part of earth's atmosphere is part of the enigma. We now realize, however, that because of our sensitive dependence on initial conditions at the beginning of time, without that antecedent act of giving, there would have been no life for us to receive from our biological parents. Attempting to trace its origins back through evolutionary time to a point or place before the emergence of our species allows me to name it a 'preoriginal' gift event: one preoriginal to human history (Primavesi 2007: 217–32).

However given, or by whom, it is important to assume (with Stăniloae) that that prior act of giving within earth-space was freely done, and that its reception too was characterized by freedom. This is presumed in Jesus' instruction to his disciples, '*Freely you have received, freely give*' (Matthew 10: 8). While the concept and the science behind what I have called a preoriginal act of giving would have been unknown to Jesus, the character of our response to the process originating from it was enormously significant for him and for those who would follow him. His injunction to them to receive and to give freely is without literary parallel in the biblical records. But neither its brevity nor its lack of literal support should be taken as a sign of its unimportance. On the contrary, as we shall see later, it is a defining mark of relationships in his own life and in those of his followers.

The dynamics of giving

In the actual process of giving, the dynamic disequilibrium between giver and receiver – for the giver would necessarily have something not possessed by the receiver – facilitates and drives the process in which the receiver becomes a giver. And so the giving proceeds and evolves. It does not stop at what is given or received but endures in the relationships *between* giver, gift and receiver, even when one or all of them disappear from view. As, too, the freedom associated with all those involved endures in the process.

To explore the dynamics of the mystery of giving successfully, we must cast off what Blake called 'the mind-forged manacles' of the reigning culture in which commodity exchanges pass for gifts. We must put aside any idea of a gift being simply an object or commodity that is exchanged between two people. Or that we can hope to explain it, understand it, or calculate its value as an object that exists independently of the process of giving and receiving. Or that it can exist independent of communities of living beings involved directly or indirectly in that process: one rooted in and perpetuated through the givenness of life on earth.

So even though the most commonly perceived context for giving includes a giver, an object to be given and a receiver, you can, says Jean-Luc Marion, bracket or put aside at least one or even two of those three features. We may, for example, give a recognizable gift (of money) to a humanitarian organization (and get tax relief on it). But the actual receiver may, and probably does, remain personally unknown to us and we to him. We can even imagine, Marion says, instances where we do not know, at the time of giving, who the nominal receiver is or may be. Nor does she know us.

Marion takes as example the parable of the Last Judgment in Matthew 25 where, when someone gives food, drink or clothing to (anonymous) poor people, in effect, we are told, they have given it to Jesus. They have given their gift to a receiver they will never meet in this life. Most mysteriously of all, it is not until the end of the world that they are able even to imagine that it was given directly to Jesus (Derrida and Marion 1999: 62).

As I have said, however, there is one defining moment in the process of giving: that in which one individual gives something to another, whether known or unknown. Giving by one necessarily precedes receiving by the other, and both establishes and affects the relationship between giver, gift and receiver at a particular time. The initiating moment here is when the mystery itself is 'taken in hand' by a giver and passed on to me, and then to you, the reader. Receiving it means discerning that particular gift event as belonging to a continuum of relationships that precedes and stretches far beyond present timescales, and, practically as well as imaginatively, beyond any ability to envisage its final receiver(s).

Derrida describes this diffuse and diffused relationship stretching back beyond and between present giver, gift and receiver as 'a living thread' (Derrida and Marion 1999: 60). His image expresses the truth that anything given to one person is also entwined with others' lives: with those peripheral to, but attached in some way to the immediate circle of present donation and reception. However tenuously, however unwittingly, the gift, with its effects, is woven into and through lives unknown and undreamt of by the original giver and receiver. I do not know how what you read here will transform your thoughts, imagination and reaction to climate change. Your freedom to react as you will shows that unpredictability is part of the givenness of our theological legacy also.

From an ecological perspective this living thread of relationship is continually spun from earth's materials by a myriad of givers and receivers who are themselves bound into complex networks of mutualistic, asymmetric interactions (Thompson 2006: 372–3). Under pressure from time and gravity, the products of these interactive relationships are made manifest in geological patterns fabricated from shell, stone and fossil. As they gradually shrink and shred, they give way to, and are received by, micro-organisms in the atmosphere, the soil and the sea. So the creative process of giving and receiving continues to produce new relational configurations.

However described – and whether or not we perceive it – that creative process binds the relationships within a present gift event to those of preceding ones where the role of receiver was played by the present giver, and connects those past relationships to our present experience of, and response to, the process of giving. Paul highlights this givenness. 'What have you that you have not received? And if then you have received it, why do you boast as if it were not a gift?' (1 Corinthians 4: 7).

This insight into the givenness of life reminds us first of all that the past and present products of the process of giving provide the means and conditions for others to act as receivers and givers in the future. And second, that boasting about its present products as if we were their sole producers and they are our possessions is not an appropriate or correct response to them. For what will the next generation receive other than what we in this generation pass on to them?

From this perspective the current climate change crisis signals a systemic failure on our part to see ourselves as products of past, present and (potentially) future gift events, and therefore to realize that we provide the initial conditions that make life-giving and receiving possible for future generations. The life-enhancing threads that attach us to all those who came before and should attach their gifts, through us, to all those who come after are being broken and disconnected. Contemporary accounts of the potential 'tipping points' in the loss of biodiversity, the accelerating destruction of rainforests and the exhaustion and polluting of underground freshwater aquifers indicate that we shall not have to wait until the end of the world to imagine the effects of this failure. We shall see them in our lifetime.

Part of the reason for this failure, as I have pointed out already, lies in the ambiguous nature of gift itself, an ambiguity preserved in different Germanic languages where the German word '*Gift*' once meant both 'present' and 'poison'. The oxygen first released into the atmosphere was in itself toxic, but the interaction between it and other elements was, and is, life-giving. This emphasizes the point already made: that a gift does not exist in some relation-free zone but signifies some element in the relationship between giver and receiver. It also reminds us that we transform, combine and alter what we receive and that we are free to do so in one way rather than another.

Therefore, in inter-human gift relationships the risk or benefit to the receiver is, in the first instance anyway, in direct proportion to the donor's goodwill. This is true whether the gift takes the form of words, images, objects or proposed relationships. The double possibility survives in English usage in the two opposed meanings of 'to take something to someone' and 'to take something from someone'. So 'to give' and 'to take' are notions that are organically linked, with their meaning often decided by an asymmetry in power that favours the donor. She can take away what she has given to someone financially, emotionally or physically dependent on her.

Going back to Derrida's image of gift as 'a living thread', we can now see not only that its brightly coloured threads are spinning new life but that these are always interwoven with darker ones of death, pain and loss. Together they form the warp and weft of every life. For the mystery of giving encompasses the gift of death as well as life, and bestows impartially with them their continuing effects on present recipients who are donors of the future. Contrary to the Greco-Roman myth of the Fates, in which Clotho spins the thread of life into intermingled bright and dark lines while her sister Lachesis twists it and Atropos cuts it short, death remains an essential element in the whole fabric of life. This truth is inscribed for all time in the accumulating fossil record.

The Christian experience of death

This interweaving of life, gift and death is, however, almost totally obscured in Western Christian theological records. I say almost, for it is clearly written and proclaimed (at many funeral services) in the lines from Ecclesiastes quoted earlier:

> For everything there is a season, and a time for every matter under heaven:
> a time to be born, and a time to die;
> a time to plant, and a time to pluck up what is planted;
> a time to kill, and a time to heal;
> a time to break down, and a time to build up;
> a time to weep, and a time to laugh;
> a time to mourn, and a time to dance.
>
> (Ecclesiastes 3: 1–4 RSV)

But an entrenched theological tradition stemming from Augustine's interpretations of Pauline texts obscures this clear acceptance of death as natural. Instead Christians are taught that death (like work) is a punishment meted out by God to all living creatures because of the sin of the first man, Adam. Confronting and changing this Christian image of a punitive God is an essential part of facing the inconvenient truth of the violence inherent in today's theological climate.

The first serious challenge to this punitive divine image was mounted in Augustine's lifetime by Julian of Eclanum. In *Gaia's Gift*, I summarized Elaine Pagels' detailed account of their twelve-year-long debate as follows. First, Augustine assumes that one human being, Adam, has the power to change 'the structure of the universe'. The hubris of this view of ourselves has led me to call for what I call ecological humility (Primavesi 2004: 119–28). Second, Augustine contrasts heaven with what actually happens on earth, and then uses this imaginative reconstruction to argue for heaven being our natural home while hell is situated in earth's depths.

Third, and positively, Julian insists that each of us can *choose* our moral destiny. We are not, after Adam, all destined for hell, any more than we are all, after Jesus' death, destined for heaven. We are all, though influenced by others, ultimately responsible for the choices we make. (A postscript to that today would be that we are each, to some extent, responsible for contributing, through our present lifestyle decisions, to the destructive effects of climate change. We are also free to do something, or not, about this.) Fourth, at the same time no one individual has the power to determine the destiny of everyone else for ill or for good. And fifth, that we suffer and die shows only that we are by nature (and indeed, Julian would add, by divine intent) mortal beings, '*simply one living species among others*'. In summary, neither death with its attendant sorrow nor the suffering caused by climate change is punishment for one man's sin imposed on all of us by

God. Nor, in the light of what has been said about 'work', is our labour a punitive sentence imposed on us by the same God for the same reason.

The experience of gift

The essential connection between suffering, life and death and its potency within
the mystery of giving was made clear to me in an interview with a survivor of
Auschwitz. 'Why did you survive', she was asked, 'when so many perished?' She
answered by re-living the moment of her arrival at the camp. Separated from her
family, she was forcibly stripped, shaved, showered, given a shapeless garment to
put on and then shoved into a barn-like building where she was confronted by a
terrified group similarly shaven and clad. Fright and desolation overwhelmed
her. Then a girl broke out of the group, came over and thrust a piece of bread
into her hand. 'At that moment', she said, 'I decided to live.'

This remembered gift event, in a place synonymous with suffering, evil and
death, makes several points about the nature and the role of the gift itself that
are exemplified in the relationship between giver and receiver. Potentially, the
'living thread' of gift holds and/or brings life or death to one or the other.
This ambiguity not only connects them. It also connects us to their gift relationship. Its potency persists beyond their lives and, even after their deaths,
connects them into and beyond our own, thus taking us all deep into the mystery of giving.

Against the parable in Matthew 25 quoted by Marion, those who do give
water, food, clothing or shelter to those who need it may discover its true cost
and value in this life rather than its posthumous reward. Deaths in childbirth
where the child survives, though comparatively rare in capitalist societies, are all
too common elsewhere. There we glimpse an excess, an overflow of life from the
gift of death: an excess that embraced the girl who gave, as well as the one who
received, in Auschwitz. And this excess gifts us too, though we received no bread
directly from her hand. The excess breaks through the circle of accepted economic value systems and conveys a deeper truth about what such gifts – just a
scrap of bread! – might really be worth. A life? Two lives? Our lives?

I do not know, and, given its setting, perhaps the receiver did not know either
whether or not the giver survived. But, as in the case of the woman who anointed
Jesus for burial, wherever the story is told, her giving is remembered and
enriches our lives today. It is also true that by choosing to give away the piece of
bread, she unwittingly supported Julian's argument against Augustine. We can
choose our moral destiny, are free to transform our gifts, even from within the
most horrific circumstances.

The character of gift event

The memory of this exemplary gift event, a product of the process of giving,
shows, not surprisingly, the mysterious character of the whole by being resistant to
conventional economic analyses. It exceeds constraints of time and place; of market

laws of exchange and return; of an eye for an eye or of a *quid pro quo*. It contradicts the laws of conservation of energy, is not part of an economic chain; cannot be contained within a balanced equation, understood as a commercial transaction or held in equilibrium by counterbalancing considerations. Indeed the very yardsticks that declare this gift event impossible are the condition of its possibility.

> [Such] an event is an irruption, an excess, an overflow, a gift beyond economy which tears open the closed circle of economics.
>
> (Caputo ans Scanlon 2006: 4)

Within the community of life on earth just such an irruption of abundant and available flower seeds brought about a flourishing of animal life and reciprocal, beneficial relationships between birds, insects, animals and plants. These contributed to the evolution of the biosphere in ways that later gave emerging human communities the water, temperature range and nutrients necessary to support life. Colin Tudge puts this gift in its contemporary context. 'Our debt to trees is absolute' (Tudge 2005: 76f.).

Caputo remarks that as soon as a donor actually gives someone a gift, that puts the recipient in debt and makes the donor look good, thereby taking from the recipient and adding to the donor, which is the opposite of what (according to economics) the gift was supposed to do (Caputo 1999: 4). This uncommon perception of 'indebtedness' may lie behind the received wisdom that the one who gives is more blessed than the one who receives. And it operates too in the way the Auschwitz survivor's story was told by her.

But our common religious and cultural attitudes to earth show that we make one glaring exception to it. Do we truly feel indebted to earth? Do we really consider it more blessed for what it gives us? More blessed than we are who receive from it? If so, why do we assume that the food it gives us needs to be blessed by us? Or, in religious ceremonies, blessed by a cleric? Ought we not simply acknowledge its blessedness by giving away our thanks to flowering trees and shrubs for what they have transformed through their labour, then given and continue to give freely to us?

Does the bread on our plate, the water we drink or the breath we take make earth's gifts look 'good' to us? Or do we not take all their goodness to ourselves and attribute it to *our* labours? This shows that we treat these gifts as having no intrinsic worth in God's eyes unless we bestow a blessing on them. Or make a profit from them. But are they not, like us, products of the mystery of giving, of the divine economy that, according to Paul, relates God with earth and God with us?

Earth's gifts

Here again, however, Augustine's view of death as punishment for human sin muddies our view of earth by including it in that punishment. And whereas we are presumed to be (potentially) saved by Jesus, in the Augustinian scheme of things earth remains unredeemed, at best contributing to our salvation but powerless to effect its own. As a result, as I concluded in *Gaia's Gift,* what the earth community has given and continues to give to us through countless visible and invisible life forms, and what this means for us and for our existence, is imaginatively, religiously and literally discounted (Primavesi 2003: 134).

And so we routinely confine the mystery of earth's giving within the closed circle of humanly conceived and contained economics, of human debtors and creditors. We do not see ourselves as belonging within a continuum where members of the earth community gift us with existence: a continuum that stretches back before the emergence of flowering trees and plants and long before our emergence as a species.

For the radical truth about us is that we are all products of the process of giving. Another poet, Gary Snyder, reminds us bluntly that indeed we could not *be,* could not exist at all, were it not for the planet that has given us our very shape.

> Two conditions – gravity and a livable temperature range between freezing and boiling – have given us fluids and flesh. The trees we climb and the ground we walk on have given us five fingers and toes. The *place* (from the root *plat* broad, spreading, flat) gave us far-seeing eyes; the streams and breezes gave us versatile tongues and whorly ears. The land gave us a stride and the lake a dive. The amazement gave us our kind of mind.
>
> (Snyder 1990: 29)

Changing our minds and our thinking about earth and ourselves is difficult: but powered by that kind of amazement, possible. Connecting this amazement with our thinking about God changes that thinking too. For we begin to understand that the mysterious relational processes that create us also create the possibility of our relating to God. And the more we learn about those processes, the clearer becomes our image of God.

The question to be looked at in some detail in the following chapter is: what image of God emerges from the mystery of giving and how does it relate to those prevailing now?

8   Changing God's Image

I give thanks
To the giver of images,
The reticent god who goes about his work
Determined to hold on to nothing.
Embarrassed at the prospect of possession ...

(Kennelly 2004: 23)

How does the impact of climate change change our image of God? And with that, change the theological climate too? Poet Brendan Kennelly progressed beyond perceiving the need for such a change in the light of its effects to envisioning an alternative. For him, the need is evidenced in 'the shot bride', victim of the systemic criminal violence in modern city culture and in the 'demented soldier', representative of structural military violence inflicted indiscriminately by and in modern warfare. Implicit and often complicit in the latter is the symbolic violence of an imperial image of God: of one who possesses the power to wreak vengeance on his enemies and to punish those who disobey his commands – and who uses that power towards these ends.

Kennelly turns away from this image towards 'the mystery of giving' and to its 'giver of images'. Such mysterious giving appears almost beyond human imagining, for it is functionally different from the kind of exchange that operates within a consumerist, competitive and therefore violently possessive culture. So, against this communally sanctioned form of behaviour, he envisions a God '*embarrassed at the prospect of possession*'.

This mysterious image not only challenges that of a God defined by the possession of power over earth, its resources and its inhabitants. It should terrify us too. For we are given it at a time when, for most of us, the prospect of *dis*possession, with its accompanying states of poverty, powerlessness and dependence in all its forms, is not only embarrassing but frightening. Kennelly is also bold enough to imply that it is our desire to possess, our lack of embarrassment in accumulating possessions, that embarrasses God.

Diogenes living contentedly in his barrel would not be embarrassed before such a God. Neither would Francis of Assisi, who loved, not feared, 'the Lady Poverty'. David Galston provides a powerful, real-life metaphor for such an encounter between images of the divine when describing a meeting between Francis and Pope Innocent III. Having initially refused to see him when he came to Rome seeking recognition for his order, the Pope, enthroned and clothed in the finest garments medieval Europe could provide, was confronted by a pauper. 'What a moment!' exclaims Galston. It contrasted the 'glory' of Christendom with its own heart of compassion; it held the deceiving wealth of the Christian hierarchy before the face of authentic practice.

> In other words, it was yet another moment where Christianity could not understand the contradiction it had become and the failure it represents.
>
> (Galston 2007: 1)

Christendom continues to represent this failure as long as its imperialist character makes it incapable of representing Kennelly's reticent God: one who literally does not capitalize on his name but holds back from the power presumed attached to its possession. And by analogy, to possessions themselves. It also fails to represent Jesus. For aversion to that power is evident in his admonitions about possessions: don't acquire them (Matthew 6: 19); sell them (Luke 12: 33, Gospel of Thomas 76). And if you sell them, give the proceeds to the poor (Mark 10: 21). In regard to the last injunction, it forms part of the story of the rich young man whom, we are told, Jesus loved 'at first sight'. But the young man, stunned by Jesus' advice, goes away dejected: 'for he had great possessions' (Matthew 19: 22).

Against the background of these Gospel accounts of Jesus' attitude to the power of possessions, Kennelly's image of a God *determined to hold on to nothing* is at once profoundly Christian and profoundly counter-cultural. A positive cultural response to this image would include a determination to acquire and consume less of the 'goods' that are forced on our attention by advertising and media representations of what constitutes a 'good' life. It would also focus attention on the fact, noted in the preceding chapter, that we receive freely soil and water from the earth and energy from the sun: 'give-aways' that constitute our life support systems.

They, however, have been transformed into possessions by a technocentric mindset focused on maximizing monetary gain: one that regards nature as an infrastructure to be adjusted to our desires and used as the raw material of our technologies. Her gifts now fuel industries whose divisible benefits have become the possession of and source of profit for private and corporate shareholders. The indivisible benefits, those things we hold in common such as clean air and uncontaminated water, are less and less available to those who cannot pay for them (Franklin 1990: 116–18). This situation is worsening as climate change progresses, resources become scarcer and more technology is required to process them, making them even more expensive. Meanwhile those who can pay are urged to use more and more.

A theological response would include a determination to go about the work of dealing with climate change by dispossessing ourselves of the violent images of punitive and retributive divine power that dominate our religious culture and legitimate oppressive social structures. This would initiate a paradigm shift within our theological world view, a move away from any assumption of a divinely sanctioned entitlement to possess and use earth's resources solely for our benefit and a move towards an understanding of the effects of such presumption.

✗We would begin to see the exercise of such a presumed entitlement as a primary agent of destructive changes in climate. We would feel its challenge to the images of conspicuous possessions, conspicuous consumption and conspicuous waste that drive the engines of desire in a capitalist culture. Gradually they would be displaced by images of receiving and giving that stand outside any kind of commercial or monetary exchange. As this shift in perspective occurs, the contrast between Kennelly's image of God and the one presupposed by a capitalist culture built on 'sustained', that is, never-ending growth in consumption, stands out ever more starkly. And the challenge to theologians to discern and proclaim that contrast becomes ever stronger.

The power of images

As the paradigm changes, the mystery becomes that of how we ever came to accept such violent, capitalist images of the God of Jesus. They are manifestly inadequate for expressing what we would want to imagine as the whole truth about him, about ourselves or about God. Rather too often they better express partial truths about ourselves; in particular about those of us, still overwhelmingly men, who exercise or desire to exercise economic, military or ecclesiastical power.

Some cultural and theological climates are more favourable than others for persuasively expressing those truths. The image of God, of 'Man' and of their relationship to each other and to the earth presented to us in the Noah story, for example, emerged from a patriarchal, pastoral, tribal human society in which possessing the necessities of life was paramount; and that possession was often marked by violence. But these human societies were so small, so technologically primitive and widely scattered that their impact on the whole earth community was negligible.

Nevertheless, the stories about them, rather than those about Jesus, have assumed a mythic importance in the lives of Christians that we must take seriously. Biblical scholar and founder of the Jesus Seminar Robert Funk recounts how he suggested to a colleague of his that during a pastors' school in Northwest America he should ask them to find out from their parishioners what stories they lived by. He asked them specifically to conduct a survey to establish whether any part of the biblical story was still alive as myth, and if so, what part. In the weeks that followed the reports he received were nothing short of astonishing.

They showed that Protestants in the northwest do have a living biblical myth. It is the epic of the Exodus and the settlement in Canaan. What is more, they associate that story with their own frequent excursions into the wilderness on weekends and holidays: they believe themselves the better for living on the edge of a wilderness, a frontier, and they think that accounts, in large part, for their piety:

> Moreover, they think God commissioned Americans to come to this continent and found a new society, a city set on a hill for all to see and imitate. But these same Protestant parishioners do not have a Christian story, that is, they have no story of which Jesus is a part. Their story is essentially the Hebrew epic.
>
> (Funk 1990: 1–4)

This shows the enduring power of the Exodus story in sanctioning colonization and in the identity formation of landless people: even when they are living securely on their own land. It also brings us back to the question posed in Chapter 6: 'Who are the Canaanites today?' The geographical and social climate in which the biblical stories are told and retold may have changed radically over millennia. The theological climate, however, has, to a large extent, remained the same. This marks our failure to recognize the pervasive violence at the heart of biblical story lines and images of God.

> The 'inconvenient truth' is that violence-of-God traditions, including violent images of God, dominate the Bible and the Quran. A corollary truth is that when Jews, Christians or Muslims use religion to justify violence they can do so based on *reasonable* readings of their 'sacred' texts.
>
> (Nelson-Pallmeyer 2007: 9)

His context is the terror attacks of 11 September 2001, and the way in which they have pushed issues concerning religion and violence into public view, albeit through a limited and often distorted lens. He acknowledges that the unwillingness to fully confront this is related to fears that doing so would undermine the authority of Scripture and pose profound challenges to basic traditions. His concern is with the way in which violent images of God and violent expectations of history permeate Christian worship, theology, liturgy and songs. And so we deny that the problems of religion and violence are *rooted in the violent content of the 'sacred' texts themselves* (Nelson-Pallmeyer 2007: 9).

These problems affect not only relationships between Christians, Muslims and Jews, but between us and all living creatures. The 'sacred' story of Noah bears this out. Today we see 'the inconvenient truth' of human failure to change the *theological* climate reflected in the 'inconvenient truth' of our responsibility for global climate change and its effects on the whole community of life on earth. Both failures are reflected back to us in the 'inconvenient truth' of an unchanging, violent image of God. For as Nelson-Pallmeyer says, if religion is going to help us respond creatively to a world torn apart by injustice, engulfed in a spiral of violence and, I would add, in the throes of climate change, then we must take an honest look at and counter the violence-of-God traditions at the heart of our 'sacred' texts (Nelson-Pallmeyer 2007: 15).

In regard to the violence inherent in the claim to possess or own earth's resources, some of the consequences of this attitude for our relationships with others are articulated in a poem of Rilke's: who also points to some of its consequences for God. In this context, it is permitted and necessary to say that God is *also* a person: one who suffers the consequences of our actions (Buber 1970: 181).

Those consequences centre on the fact that our desire to claim possession of everything, from tree, to life, to wife by saying 'it's mine', creates alien and indeed alienating images of them. And of God.

Rilke counsels God against worrying about this disastrous tendency.

> You must not worry, God.
> They say: 'it's mine' of everything,
> Which suffers this in silence.
> They're as wind that lightly strokes a branch and says: MY TREE ...
> They say: my life, my wife, my dog, my child,
> Yet know full well that everything: life, wife, dog and child
> Is but an alien image
> which blindly, with outstretched hands, they stumble on.

These 'paltry ramblings', says Rilke, create no bond with anything around us. For the claim to possess others blinds us to the integrity of what they are in themselves, and to the integrity of Godself. So he implores:

> Dear God, hold on to your self-possession.
> Even those who love you and discern your face in darkness,
> like a flame trembling in your breath – do not possess you.
>
> (Rilke 1975: 48. Trs C. Carr and A. Primavesi)

By definition we never possess an adequate image of Godself, for that mysterious, transcendent integrity is forever beyond our reach. But some images are more alien to its truth than others. With this in mind, in a culture obsessed with possessions and with possessing, and in an environment ravaged by the violent effects of this obsession, we must look for images that re-envision the existing bonds between us and with all living beings around us.

This means, suggests Kennelly, listening to the sound of doors opening and closing in the street and hearing them like

> [T]he heartbeats of this creator
> Who gives everything away.
>
> (Kennelly 2004: 25)

Including, we may now suppose, the power either to initiate or to eradicate the effects of our violent obsessions: evident in greed, war, degraded environments and terror.

The traditional theological climate

To juxtapose images of a powerless creator god with those of greed, possessiveness, war, devastation and terror challenges traditional expectations of divine power. What do we, by implication at least, normally expect God to do about 'such things'? The short answer is that we hope and pray for him to destroy them, or to deal with those involved by removing them from our sight (to hell?) through his overwhelming power. This violence-of-God tradition is most clearly present today in the dominance within religious thinking of an all-powerful, wrathful, punishing image of God that can be found, for those who seek it, in the Bible. Widespread acceptance of such violent images means that although people may not kill each other over religion, violence against others is more likely when religious differences are among issues that divide us. They are a useful *ally* when we want to excuse the use of violence against others.

What kind of theological climate propagates and then sustains such acceptance of wrathful images of God and violent expectations of human history? Broadly speaking, it is one in which God, as a theological entity, is presumed to have and to use divine (as opposed to human) power to deal with the horrifying historic realities associated with war. And is expected to deal with them in ways that create ever more shocking scenarios – but this time carried out in God's name. This symbolic climate normalizes the image of a God not only capable of responding to violence with overwhelming force, but one who is expected to do so. The violence is simply redirected against those named or perceived as God's enemies.

So in the context of today's climate change crisis some Christians continue to pray that such a God will 'destroy those who destroy the earth' (Revelation 11: 18), even though doing so would mean accepting and perpetuating in all its horror the kind of murderous scenario gratefully acclaimed and assented to in the verse in Revelation preceding the one quoted above:

> We give thanks to thee, Lord God almighty, who art and who wast,
> that thou hast taken thy great power and begun to reign.
> The nations raged, but thy wrath came,
> and the time for the dead to be judged,
> for rewarding thy servants, the prophets and saints,
> and those who fear thy name, both small and great,
> and for destroying the destroyers of the earth.
>
> (Revelation 11: 17–18 RSV)

This image of an almighty, all-powerful, wrathful, punitive and destructive God is traditionally hymned and depicted in all his majesty in Christian worship and buildings. He is the reverse image of Kennelly's god: for he possesses and capitalizes on his power over all the earth and over all its inhabitants, living and dead.

The emergence of this Christian climate

The contrast between these images could not be greater. For which of them can we now, honestly, give thanks? Should we not ask who gave us the images of a wrathful, violent God and why we receive them without questioning their destructive effects? Those who coined the images in the above text from Revelation were expressing an understanding of their world dependent on their experience of earthly imperial power. Their images were shaped reactively to the rise and fall of great empires in Europe and the Middle East. But the thought-world and imaginative landscape behind that kind of 'end-time' vision recorded by them is foreign territory to most modern readers and theologians. Not surprisingly, some iconic representations of the God presiding over and passing judgment on the landscape, as in the passage above, bear a strong resemblance to those previously representing Zeus as the Olympian Father dispensing justice over all the earth.

> Then Zeus the father on high took his golden scales
> In them he put the two fates of death that cuts down all men.
>
> (Weil 2005: 13)

Today, some still seek refuge in this overriding vision of an exercise of divine power, especially when confounded by their personal inability to deal with the harshness of contemporary reality. The alternative theological reality they choose for refuge is described in literary accounts of 'revelations' (*apocalypses*) to a prophet or writer obsessed with hope for a divine irruption into a present state of affairs, usually one verging on catastrophe. Allied with this is the longing for a divine realm in which God will reign as king. This type of literature flourished at times of crisis in Judaism and Christianity. The bleak world view of the apocalyptic vision in Isaiah 24–7, Isaiah 56–66 or Zechariah 9–14 stems from the turmoil of the late sixth and fifth centuries BCE.

From the Roman persecutions of Christians in the first century CE came Mark 13, the book of Revelation and a considerable number of extracanonical writings like 4 Ezra. These were generated within communities suffering from destructive imperial violence and were circulated as a means of sustaining hope for a new and better life; if not here, then in eternity. Deliverance from the power of the Roman Empire would come through the power of a God loudly hymned today as 'mightier still' (Nelson-Pallmeyer 2001: 143–9).

Such descriptions of God arise naturally within societies whose power structures are hierarchical and patriarchal. Describing God as a totally transcendent, changeless being, one inhabiting an unearthly realm (but nevertheless always a male!) comes naturally in a culture dominated by neo-Platonism (Marion 1991: 72–83). Brought together, these influences generate (as in Augustine) descriptions of the City of God as a realm of eternal happiness; one in which necessity has no place and where there is enjoyment of a beauty that appeals to reason. The obverse of this is the earthly city whose citizens do not know 'the Founder of the

Holy City', preferring their own false gods who are 'deprived of His unchange-able and freely communicated light'. This binary code had a particular resonance when Augustine described it in CE 413. Just three years earlier the barbarians had sacked Rome.

Sustaining the climate of violence

This historic example of Christian expectation of God's punitive violence wreaking havoc on earth demonstrates the power of such hopes. They have a self-validating effect: that is, we tend to see or hear or express, from within our particular envi-ronments, whatever resonates with those expectations. Furthermore, the expectations themselves are validated because they shape our actions in ways that bring about what we believe, hope or fear may be the case. The fall of Rome vali-dated Augustine's Platonic perception of God as one who transcends all worldly order. His writings against, and his successful persecution of, those with a different expectation of God, such as Julian of Eclanum and Pelagius, appeared historically validated by the triumph of his theology over Julian's and by Pelagius being sent into exile.

I am concerned here with setting patriarchal, hierarchical thinking within his-torical contexts and showing it effective in different historical processes rather than showing it as 'wrong'. This is part of the ground-clearing exercise necessary for giving us a positive, nonviolent perspective on our relationships with others and with the earth. I say this even though theologically it can never be right to abuse God, so to speak, in order to validate abuses of power. But it is certainly right to draw attention to the continued acceptance of such abuse of God's image in a contemporary theological climate where, in spite of the rise of envi-ronmental, racist, egalitarian and feminist consciousness, Christianity remains dominated by violent patriarchal images of God that may easily be used to sanc-tion violence of all kinds.

Even where societies pride themselves on belonging to a secular state, there remains a presupposition of 'top-down' interrelationships within it that mirrors those presupposed between God and the world, and within the world. This pre-disposes us to rank social order in terms of subordination to powers 'above us' that at the same time validates the power of those with subjects 'under' them. This in turn validates a usually notional transferral of ultimate responsibility for what happens in the social order to a sovereign ruler or minister. In the case of God, one who is presumed to dwell outside as well as above that order of being.

This brings me to a crucial point about patriarchal images of God. They pre-sume that God dwells within and speaks from an unearthly realm beyond emotion, imagination or weakness. This supports the idea that He is changeless. For time and consequent change is used to define earth and its inhabitants, but not God. Therefore, since evolution is a 'change through time', God is presumed not to change or evolve in any way that relates to our own evolution, to our own experience of life and death and to the emergence of new life forms or to be sub-ject to or a subject of changing images.

But describing a relationship with God that does not allow for change, either in God, in us or in our theological descriptions, implies the death of that relationship. For to be alive is to change. Any claim to an unchanging relationship denies the possibility of growth in knowledge between those relating to each other. Does a baby relate to its mother and father in the same way as an adolescent? Yet there remains an assumption in theology and in worship that we continue to be bound by modes of address and descriptive images that are distant from present-day life yet assumed to retain their relevance. Our personal relationship with God is presumed to remain unchanged from schooldays to grave; and our communal relationship with the divine is presumed to remain unchanged from the time of Plato, Aristotle, Isaiah, Ezra and Augustine.

How would this work in reverse? What images would any of these philosophical or theological luminaries use to speak about God today? What language would they borrow to describe their relationship with the divine? Could they ignore the evidence for evolutionary change and now for the impact of the earth sciences and the study of climate change? The refusal by many church institutions even to consider changing their patriarchal imagery can be seen, ultimately, as a refusal to consider Christians as members of an evolving society affected by the evolution of knowledge.

Therefore, the inconvenient truth of violence-of-God traditions remains a major factor in the dangerous inertness of our theological climate; an inertness that, by default, sanctions violence against other members of the whole community of life on earth. If challenged, that default position is justified on the grounds, for instance, of the early Genesis narrative:

> Be fruitful and multiply and fill the earth and subdue it; and have dominion over the fish of the sea and over the birds of the air and over every living thing that moves upon the earth.
>
> (Genesis 1: 28)

This is presented as prefiguring, and so validating, the fear and dread imposed on the earth and its nonhuman inhabitants by Noah and his descendants. But if, as has proved to be the case, such a theological climate cannot sustain a healthy relationship between us and all other creatures, then the question is not *whether* we should change our images of a violent God but what other images are there that we should embrace.

Changing the theological images

What images can we give thanks for today that are already there within our Christian religious tradition? Images that would generate compassion, goodwill, generosity of mind and spirit?

There is already a vital connection between Kennelly's non-possessive, nonviolent creator god and the nakedness and powerlessness implicit in Paul's image of 'the weakness of God'. It is there too in the Pauline emphasis on God's choice of 'the things that are not' (1 Corinthians: 1: 25; 28). This, he says, is not under-stood by 'the rulers of the world' but is 'revealed to us through the Spirit' (1 Corinthians: 1: 8; 10). So 'we have received not the spirit of the world, but the Spirit which is from God, that we might *understand the gifts bestowed on us by God*' (1 Corinthians: 1: 12).

The particular gift revealed to us in this image of a weak, naked god is crucial in creating a theological climate built on and sustained by the mystery of giving. It offers us a choice between the naked powerlessness of God in Jesus and the firmly entrenched image of Christendom: that of a God cloaked in sovereign power; an almighty lord on which every terrestrial sovereignty is modelled (Caputo 2006: 32). But the question has already been raised as to whether in fact the modelling is not the other way round. Are not those earthly sovereigns them-selves the model?

Countering them, silently confronting them, Kennelly offers us instead the scandal, the stumbling block of a God who, embarrassed at the very prospect of possessing power, '*gives everything away*'; strength, possessions, life, family and friends. This is one who dispossesses himself of every hierarchical attribute: such as pre-eminent power, glory and Solomonic wisdom. The challenge of under-standing such a God, especially in the present theological climate, is, according to Paul, too great for human reason. It can only be met through receiving the Spirit of this God. Then we might understand, to some extent, the mystery of giving associated with the God who dispossesses herself of everything.

Such an image of God has resonated most deeply with 'what is weak in the world' (1 Corinthians: 1: 27); with the 'are nots' and 'have nots'; with a young girl in Auschwitz; with the poor, women and nature: with those dispossessed of their lands, disowned by an entrenched hierarchical tradition and, if considered at all, relegated to its lowest ranks. Against this is an image of God as hierarch, as an idol who presides over an order of being in which the poor, women and nature are devalued and their oppression validated.

Whenever and wherever a religious patriarchal tradition such as Christianity has been, and is, defined in terms of male sovereignty, women and nature have at best been seen and treated as bearers of 'weak' power. These unrepresented 'nobodies' have been defined as such by those 'strong' in body, voice and buying power. For the unrepresented, 'giving everything away' is not a denial of God's presence. It is an affirmation of its mysteriousness; a critique of the idols of pres-ence and their claim that anything *present* can embody God or claim to be God's definitive form (Caputo and Scanlon 1999: 5).

Caputo sees the Pauline image of God's weakness as short-circuiting the power and wisdom of the world, radically distinguishing God's power from that power, and so inspiring us to derive from it a weak, provocative force that inter-rupts the usual power supply answering to the name of God. This weak force erupts into our lives as the name of 'an event rather than an entity: of a call rather than a cause, of a provocation or a promise rather than a presence'. Kennelly's 'reticence', expressed in continual, excessive 'giving away', highlights what Caputo calls the 'evocative, provocative force sheltered by the name of God ... the weak force of a call that lacks the sheer brawn to coerce or translate what it calls for into fact'.

(Caputo 2006: 12–13)

Finally for now, such an idea of God, defined by weakness, by the determination to hold on to nothing, evokes one who cannot, does not impose conditions on giver or receiver. One who, like the sun and the rain, gives unconditionally: to just and unjust alike. This 'mad' condition of unconditionality, of taking no account (literally) of thanks or future recompense, emphasizes the radical, anarchic nature of the manner in which we are to give and receive when 'carrying out' the event that takes place under the name of the kingdom of God. The truth of this kingdom, says Caputo, is one we are called on to make come true, to realize. It is not the truth of a journal kept by eyewitnesses of magical events (Caputo 2006: 14–16).

9 The Gift Event of Jesus

Freely you have received; freely give.

(Matthew 10: 8)

In fact God asks that one give without knowing, without calculating, reck-oning or hoping; for one must give without counting, and that is what takes it outside of sense.

(Derrida 1995: 97)

How did Jesus, in realizing the truth of the kingdom of God, characterize the mystery of giving? Its defining characteristic, he said, is that it be done freely: without calculating, counting, reckoning or hoping for something in return. But there was nothing abstract about this definition for him. Such freedom defined the receiving and giving characteristic of his own life and death. He freely accepted water, for instance, from a despised Samaritan woman: one to whom he freely gave the knowledge of who he was (John 4: 7, 25). When he asked that we give without calculation or reckoning, his own responses to others provided the template for this. Indeed his life epitomized that of 'a god determined to hold on to nothing'; not even his life. For, 'embarrassed at the prospect of possession', he held on to no power with which to resist those who laid hands on him – either for their healing or ultimately, for his own death.

The kingdom of gift relationships

The expectation of some return for what one gives is, however, almost axiomatic in today's world. Therefore Jesus' injunction to his disciples lays bare the radical, anarchic nature of relationships within the kingdom of God whose truth we are called upon to realize. This is not the simple-minded, street corner anarchy that reigns in kingdoms where lawlessness and unchecked violence rule. It is one that runs contrary to the laws that generally govern relationships between givers and receivers in the kingdoms of this world. It is one where the only 'object' in view is due attention to the nature of the act of giving and to the intention to give. In the end, says Derrida, the gift or object given doesn't count: it is not taken into account (Derrida 1995: 112).

This reprises the discussion in Chapter 7 of the nature of relationships within the mystery of giving. Jean-Luc Marion, in conversation with Derrida on the question of 'what is the gift?', agreed that we cannot explain it, nor have access to its meaning if we keep it within the horizon of the world's economy. Indeed we could, he said, put aside at least one or even two of the most abstract and common patterns of gift-giving: a giver, a receiver and an object given. For there are situations where the giver is absent, or the receiver is unknown or when nothing, no-*thing* is given. This happens when we give time, when we give our life, when we give death or when someone is given power to act (Derrida and Marion 1999: 62–3).

Derrida agreed that a gift does not belong within a series of objects. It is rather, he said, a 'living thread'. In Chapter 3 I viewed this thread from an ecological perspective: as one spun from earth's materials by a myriad of givers and receivers who are themselves bound into complex networks through mutualistic, asymmetric interrelationships. Now I want to view it from a theological perspective: weaving it into and through more familiar, but no less essential patterns of relationships between receivers and givers within the mystery of the kingdom of God.

For there are kingdoms, says Caputo, and there are kingdoms. A 'kingdom' is nothing in itself apart from its historic context, its natural environment and the web of relationships that hold it together. Within the context and environment of Jesus' own life, death and teaching, we see that what reigns over the relationships within his context and environment is the sovereignty of the Roman Empire, maintained through force or terror and sustained through the use of overwhelming power. Within God's kingdom, however, things are organized around the 'weak' power of the powerless; freely shared among the dispossessed. Those within this kingdom are therefore ruled by an unconditional summons to stay open to the call of justice, to the claim or appeal of those without force. It is a justice, Caputo speculates rather wildly, that would leave the ninety-nine with which the law is concerned 'while it goes off in search of the missing one!' (Caputo 2006: 27).

The exclamation of astonishment accompanying his speculation reminded me of a rabbinic story about what would now be called the 'leadership style' of Moses.

> While Moses was feeding the flock of his father-in-law in the wilderness, a young kid ran away. Moses followed it until it reached a ravine where it had found a well to drink from. When Moses reached it, he said, 'I did not know that you ran away because you were thirsty. Now you must be weary.' He carried the kid back. Then God said, 'Because you have shown pity in leading back one of the flock belonging to a man, you shall lead *my* flock, Israel.
> (Montefiore 1974: 45)

When God looks for someone to lead his people from oppression to freedom, say the rabbis, he looks for someone who is consistently compassionate; someone whose attentiveness to the world and especially to its smallest, weakest and most lawless creature can be taken as guarantee of the attentiveness needed for the great task of leading Israel.

Such thinking about who should be given power over others presents us with a model based on recognition of the interrelatedness of all being: whether our own being before God or our being in the world. This kingdom is the product of a certain kind of relationship between a community of subjects and not the sum total of a collection of objects (Berry 2006: 8). Because whatever we do to the least powerful among us, to the smallest and apparently most insignificant member of the community of life on earth, ultimately affects all we are and do to each other.

That lesson is presently being driven home through increased understanding of the effects on our life community of the presence or absence of ocean algae, of subvisible and intangible organisms in the air and those visible on earth but discounted – such as bees, beetles and earthworms. From such as these we receive the gifts of clean air, rain and fertile soil, flowering plants and their fruits. For the rabbis, the wholeness of the earth community is determined primarily by the actuality, by the oneness of all God's relationships with the world. The God of the thirsty kid is the God of Moses.

Among the many stories recorded of Jesus' attendance at synagogue there is a very short one, found only in Luke's Gospel, where he notices a woman in the group, bent over, unable to stand upright and praise God with everyone else. In contrast to other healing stories, she makes no request to be healed. Jesus, however, attentive to her plight, calls her and lays hands on her. She is given the power to straighten herself, lift up her head and praise God.

The leader of the synagogue objects that it was not necessary to break the Sabbath law: there are six other days in which she could come to be healed. Jesus does not argue with him on the basis of saving human life. Rather, he says, just as it is permitted to care for household animals on the Sabbath, so can one act towards a fellow human being. As in the rabbinic story, the point of comparison for one's conduct towards others is one's behaviour toward animals (Luke 13: 10–17). And in each case, the receiver of the gift of attentive care is a member of the community of the kingdom of God powerless to find care for herself. In the latter instance, one whose need is ignored or discounted by the ruling hierarchy in the name of God – but in fact in favour of keeping man-made laws.

The climate of the kingdom

Martin Buber extends this logic when he says that the bright edifice of community, built (in the rabbinic story) upon the relationships between Moses and the needs of his animal and his human 'flock', cannot be raised other than on a foundation in which our attentiveness to God is indistinguishable from our attentiveness to the world (Buber 1970: 155–7). This creates a theological climate in which violence against any living being, the use of force against those who do not conform to accepted rules of conduct or the use of religious sanctions to legitimize a lack of compassion is automatically ruled out.

For this is the climate that sustains a kingdom whose laws are organized around the needs of the weak, the have-nots and the powerless: epitomized here by a young animal and a disabled woman. Its laws embody concern for those excluded from the benefits of the laws of this world and support unconditional giving as a way of responding to their needs. Observing these laws, rather than those ruled by the economic interests of property owners (represented by Moses' father-in-law) or by those with power to exercise religious sanctions (represented by the synagogue ruler) means that the God of this kingdom dispossesses himself of the kind of power that rules 'outside' his kingdom.

This means, says Derrida, that the laws of his kingdom take us 'outside of sense': outside of what makes sense according to the reckoning of reigning cultures. Outside, that is, the usual expectations of gift returns within the present world economy or the prolongation of suffering in God's name. The climate of this kingdom sustains a different kind of theological environment to that which has evolved around a God of power, might and punitive violence.

The God of this kingdom

Given this difference, encapsulated in Jesus' unconditional demand to receive and to give 'freely', we then have to ask: what kind of God reigns within a kingdom whose relationships exemplify the mystery of giving? What, in particular, can we properly expect of the God who gives us the gift of Jesus? How do the 'usual rules' of giving apply here? For they emerge not from the economic systems of this world but from that *economy of the mystery of God* through which, we are told, God so loved the world that he gave us his only Son (Ephesians 3: 9; John 3: 16). And it is this love, given without reserve, that is at the heart of the mystery of giving within the kingdom of God.

However (and however understandably), even as he proclaims the mystery of divine love manifest in the gift of Jesus, the evangelist proceeds to make 'sense' out of it. He does so by claiming that its intended effect is the gift of eternal life to those who believe in Jesus: 'God so loved the world that he gave his only Son *so that whoever believes in him should not perish but have eternal life* (John 3:16). This reason is given in spite of Jesus' insistence that *our* gifts are to be unconditioned and unconditional. Surely then, we could not expect less of God than Jesus expects of us! The prior condition of our receiving his Son freely is that divine gifts too be unconditional: that Jesus has been freely given to us by God. It is *because* the gift of his presence is free, without our being able to earn it or to calculate its worth – our salvation or thirty pieces of silver? – that Jesus demands we give freely to others.

But this kind of giving is 'outside sense'; it does not appeal to reason. Therefore, we have been ready, indeed eager to believe the evangelist when, with Augustine as his most influential theological successor, he 'makes sense' of the gift of Jesus by introducing a punitive element as latent in the concept of gift-giving. The conditional divine punishment accompanying the gift of Jesus potentially operates on two levels. First, the gift itself is given a twofold purpose or intention: that those of us who do believe in Jesus shall not perish but will have eternal life. Conversely, those who do *not* believe in him *will* perish.

Second, and most obviously, Jesus' own living out of the mystery of the kingdom of God ends in a cruel and tortured death inflicted through the routine sovereign punitive power of the Roman Empire. Yet that death is presented to us *theologically* as God's instrument: as the price exacted by God for our deliverance from death. His death is presented as the price of 'giving' us 'eternal life'. At what cost!

The gift of Jesus

This theological 'consensus' about the gift of Jesus keeps the nature of that gift hidden behind the safe, rationally constructed walls of a theology of salvation. Seen from behind them, those who do *not* believe in him are 'condemned' by the wrath of God, that is, not 'saved' from it (John 3: 17). From behind these barriers we can no longer discern what lies beyond them: the 'senselessness', foolishness and mystery of a God who freely gave us the gift of Jesus – with no heavenly strings attached. Staying behind them, we potentially deprive God of the very freedom in giving us the gift of Jesus that he, in response, wants his followers to exercise in receiving him freely. For it is this very condition (of unconditionality) that makes Jesus the gift of a divine and not a human economy: of *the economy of the mystery hidden in God* (Ephesians 3: 9).

Within the theological climate of a human-centred, economically reasonable salvation theology, emerging naturally in the environment of an all-powerful, punitive and retributive God, the mystery of divine gratuitousness and incalculable givenness in the exemplary gift of Jesus is hidden from our view. As is the free gift of the death of Jesus – given us freely, unconditionally by him and not as payment to an angry God.

The kingdom preached by Jesus is one where all relational transactions, above all those between God and ourselves, are marked by freedom between giver and receiver: and not by the coercive power that has historically led us to forcibly baptize others in the name of Jesus and/or make them bend the knee at the sound of that name. Freedom to receive, or to reject the gift of Jesus, is precisely what makes the kingdom he preached the kingdom of God – and not a kingdom of this world.

Therefore, the nature of relationships within it are spelt out by Jesus as being unconditional, nonviolent and based on love without reserve. All of which is manifest in his own life and teaching. At the individual level it requires of us the willingness to give everything away that the world holds dear: to dispossess oneself of all power, hatred and desire for remuneration of some kind.

The nature of this kingdom

Caputo follows through on this 'logic' in order to point out that we are speaking here of the possibility of a kingdom of the kingdom-less, one where there is no sovereign and no one reigns – or if they do, they have no power. It is a kingdom where the only kind of power that is permitted is powerlessness, where the very condition of power is that it is without power:

> When the Roman soldiers mocked the so-called 'king of the Jews', telling him to come down from the cross, the irony was instead visited on them. To speak of a 'kingdom' in a case like this would indeed be an irony, but one in which a mighty kingdom like the Roman Empire was being mocked by declaring a bedraggled bunch of low-born and powerless people a 'kingdom'.
> (Caputo 2006: 26)

It is worth reflecting here on the dialogue recorded in John's Gospel between Jesus and Rome's representative, Pilate. In response to the latter's question: 'Are you the King of the Jews?' Jesus replies that his kingdom is not of this world, for if it were, his servants would fight to prevent him being handed over to the Jews. 'But my kingship is not from this world' (John 18: 33f.). John's Gospel is generally accepted as being written later than the others. In this lengthy dialogue (as in the statement in 3:16f. mentioned above) one can hear the efforts made long after Jesus' death to discern, understand and come to terms with the nature of his kingdom: as opposed to our experience of any earthly one. Its inclusion in a 'gospel', that is, as 'good news', means that it typifies a 'good' kingdom from Jesus' perspective – but not from a Roman one (Crossan 1999: 285, 288–9).

Its disparity with any other kingdom is highlighted, as we have seen, in the passage in Matthew's Gospel, where, sending his disciples to preach the 'at handness' of his kingdom, Jesus commands them, and all within it: 'Freely you have received, freely give' (Matthew 10: 8). The lesson is driven home when the verses following this injunction set out some of its more unusual consequences: to carry no gold, silver or copper coins; no knapsack, second shirt, sandals or staff. As a consequence, they are dispossessed of the usual accoutrements of economic, cultural and physical power.

By setting these criteria Jesus creates a theological climate for them to work in that is totally at odds with the prevailing one and its culture of imperial power and possession. Instead of the usual 'terms and conditions' for a relationship with God he gives his disciples an image of one defined by what that world sees as weakness, as powerlessness, as dispossession. But then, as Simone Weil says in her reflections on the use of force in accounts of the Passion of Jesus, 'only he who has measured the dominion of force, and knows how not to respect it, is capable of love and justice' (Weil 2005: 35).

Loving and giving

At the end of Chapter 5 of Matthew's Gospel Jesus had established the character of his God as one who loves without reserve: exemplified in the command to love one's enemies. Commenting on this, Derrida remarks that if you love only those who love you and to the extent that they love you, if you hold strictly to this symmetry of mutuality and reciprocity, then you give nothing (Derrida 1995: 106).

This theological perspective on our relationships is completely at odds with a world view in which an enemy is essentially defined as an object of justified hate, worthy only of violence, death or rejection. But such an apocalyptic vision fits well, alas, with the image of a wrathful, punitive God.

From 'outside' that theological climate, however, Jesus consciously offers us an image of divine love without reserve – and therefore without violence:

> *You have heard that it was said,* 'You shall love your neighbour and hate your enemy.' *But I say to you,* Love your enemies, bless them that curse you, do good to them that hate you and pray for those who persecute you, so that you may be children of your Father in heaven. *Because* he makes the sun rise on the evil and on the good and sends rain on the just and the unjust. For if you love them that love you, what reward have you? Do not even tax-collectors do the same? And if you salute only your brethren, what more are you doing than others?
>
> (Matthew: 5: 43–6) [Italics mine]

This 'mad' condition of unconditionality, of taking no account (literally) of thanks or future recompense, emphasizes the radical, anarchic nature of the manner in which we are to live out the mystery of giving; to 'carry out' the event that takes place under the name of the kingdom of God.

Here the theological weather systems do not discriminate in favour of the good or deprive those who are unjust of their necessary share of sunshine and rain. Its divine economic system leaves us nothing to bargain with: nothing to give us an advantage in 'free' market exchanges. All of which bears witness to Caputo's 'madness' of the kingdom of God (Caputo and Scanlon 1999: 6). For if this is the kingdom that is at hand, that hand is empty, soft and vulnerable as that of a child. It is not the sort the world is used to seeing or grasping (Caputo 2006: 26).

Its paradoxical nature is expressed for me in a translation by Jean Bollkaemper of a poem by Erich Fried. Her friends received it as she intended: as her epitaph,

> It's nonsense, says reason.
> It is what it is, says love.
> It's a disaster, says logic,
> It's nothing but pain, says fear.
> It's hopeless, says common sense.
> It is what it is, says love.
> It's ridiculous, says pride.
> It's foolhardy, says prudence.
> It's impossible, says experience.
> It is what it is, says love.

Within this theological realm, where the weak force of its God is defined by its opposition to, and subversion of, the sovereignty and established order of Roman rule and philosophical reason, we can appreciate the deliberate play on the word 'passion' by New Testament scholars Marcus Borg and Dominic Crossan. The term 'passion', they say, in the context of its traditional Christian background, is from the Latin noun *passio,* meaning 'suffering'. But in everyday English we also use 'passion' for any consuming interest, dedicated enthusiasm or concentrated commitment. In this sense, a person's passion is what she or he is passionate about.

The first passion of Jesus, they say, was the kingdom of God, namely, to incarnate the justice of God by demanding for all a fair share of a world belonging to and ruled over by the convenantal God of Israel:

> It was that first passion for God's distributive justice that led inevitably to the second passion by Pilate's punitive justice. Before Jesus, after Jesus, and, for Christians, archetypically in Jesus, those who live for nonviolent justice die all too often from violent injustice.
>
> (Borg and Crossan 2007: viii)

This historic setting for Jesus' preaching of the kingdom relocates the mystery of giving, and the gift of Jesus as exemplar of that mystery, within what William James calls 'the pragmatic way' of taking religion that is also 'the deeper way':

> It gives it body as well as soul, it makes it claim, as everything real must claim, some characteristic realm of fact as its very own.
>
> (James 1960: 493)

What could be more real than Jesus' body, his mortal and mortally wounded body? Crossan draws attention to its power. For Christians that is the historical gift that connects Jesus-then with Jesus-now as risen Jesus. There is, he says, ever and always, only this one Jesus for Christians. The test is this: does the risen Jesus still carry the wounds of crucifixion? In Christian gospel, art and mysticism, the answer is clearly yes. Those wounds are the marks of the historical gift of Jesus. To understand them, you need to know about his death. But to understand the death, you would have to know about his life, reconstruct its history with the fullest integrity and then say and live what that reconstruction means for present life in this world (Crossan 1999: 302, 307).

Here it means that through the freely given gift of Jesus' body, we are freely received into one theological 'kingdom' ruled only by the indiscriminate, unreserved love of God. It relates us with Jesus-then and with Jesus-now within one community of life on earth organized around a divine economy that takes particular account of the smallest, weakest and most needy of its members. And it binds present gift events to preceding ones in a continuum stretching back beyond human history in which the role of receiver was played by the present giver.

For what did Jesus have that he had not received? And what do we have that we have not received? Including the free gift of Jesus-then and now.

10 The God of Jesus – or of Caesar?

What sort of God would one believe in if this God were to be seen in Jesus' words and deeds?

(Patterson 1998: 10)

Why do you call me good? No one is good but God alone.

(Mark 10: 18; Luke 18: 18–19)

When a rich man greets him as 'Good teacher' Jesus refuses to be addressed as 'good' on the grounds that God alone is worthy of this description (Luke 18: 17–18). Their exchange underlines not only the 'indescribability' of the gift of Jesus discussed in the previous chapter. It also reflects the Jewishness of Jesus in claiming the quality of goodness for God alone (Funk 1993: 91). And as God by definition defies all description, divine goodness necessarily exceeds any human demonstration of it and goes beyond any criterion we might employ in attributing it to others.

Yet in spite of Jesus' attempt to focus attention on God rather than on himself, it is reasonable to presume that the young man may have heard Jesus described as 'good'. Why else greet him publicly in this manner? Incidents like this, recorded in early traditions about the person and life of Jesus, give us a glimpse of how he was perceived by others. They also build up a picture of God that can be attributed to Jesus. When addressing him in prayer as 'Father!' Jesus appears to have made a conceptual transition from an awareness of God as compassionate creator to seeing God's relationship with humans as symbolically represented in a paternal role (Weeden 2007: 82). This presupposes qualities of love, compassion and care associated with the best of human parents. It also resonates with the rabbinic image cited in the previous chapter of a consistently compassionate shepherd and leader and with the feminine image, attributed to Jesus, of Wisdom gathering her children in her arms, while weeping over the violence inflicted on them, as a hen gathers her chicks under her wings (Matthew 23: 37–9).

Why then have such nonviolent divine images attributable to Jesus been largely ignored by Christians in favour of those in the Bible that present God to us as a wrathful, punitive judge? And indeed, depict Jesus himself as an awesome figure surrounded by emblems of judicial power? Although not addressing this question directly, James Robinson helps provide a possible answer when he remarks that the topic of 'the real Jesus', the real historical individual, did not even exist until the Enlightenment; unless one wanted, as a latter-day Monophysite or Arian or Adoptionist, to revive long since forgotten heresies. It was only with the historicism of the nineteenth century that interest in who Jesus really was and, by implication, the import of what he said and did, emerged: that is, emerged as a question that could be answered through objective historical research (Robinson 2007: 65). And it is only since then, working with and through that research, that questions about 'The God of Jesus' could be posed – or answered (Patterson 1998: 27–45).

Therefore, says Robinson, up until modern times people could only know about Jesus through their religious experience in the Church, codified in doctrines based, not on a text Jesus taught his disciples and remembered by them, but rather on creeds that were projected back onto Jesus' life and death, such as the Apostles' Creed, which developed out of the second-century baptismal liturgy of Rome (Robinson 2007: 237).

> But in that familiar creed, Jesus' own history, what he himself said and did during his life time, is fully bypassed. Not what he said and did, but only what they said about him, counted as saving information: born of the virgin Mary, suffered under Pontius Pilate. But what lies in between? Is that of no significance?.
>
> (Robinson 2007: 82)

Robinson attributes the general lack of knowledge, indeed of curiosity, about who Jesus really was, to the fact (mentioned already in other contexts) that Church hierarchies have told us, with an imperious authority before which all doubts were thrust aside, that all we need to know about Jesus is that he saved us from sin and from hell, the punishment for sin. That remains the primary focus of Church liturgy, teaching, preaching and practice and, therefore, the sole significance of his life and death for many Christians.

This means, as Robinson points out, that despite the fact that Paul never met Jesus, the Pauline heraldic proclamation (*kerygma*) of cross and resurrection is assumed to say everything that we could want or need to know about him:

> For I delivered to you as of first importance what I also received, that Christ died for our sins in accordance with the scriptures, that he was buried, that he was raised on the third day in accordance with the scriptures, and that he appeared to Cephas (Peter), then to the twelve.
>
> (1 Corinthians 15: 3–5. See also Romans 1: 1–4; Thessalonians 1: 9–10)

This proclamation not only represents the process by which faith in Jesus was assuming a hard-and-fast outline by the end of the first century CE but also testifies to the fact that the process was at work at an even earlier stage. For Paul exhorts his correspondents in Thessalonia to 'hold fast to the traditions which you have been taught', those same traditions that, as he tells the Corinthians, he himself had received (2 Thessalonians 2: 15; 1 Corinthians 11: 23; 15: 3). These lie behind the Apostles' Creed and all subsequent creedal formulas. They summarize Jesus' 'work' as a whole, setting it within the established Jewish context of prophetic yearning for the coming of the Messiah, the one who would deliver Israel from her enemies. They give supreme significance to the catastrophic fact and manner of Jesus' death by seeing it as our personal deliverance from the punishment of sin (Kelly 1950: 9–48).

That interpretation of Jesus' crucifixion by the Romans became the focus of Church teaching, was enshrined in ecclesiastical architecture and visual art and reinforced in hymns, sermons, catechetics, religious calendars and artefacts. These, by their very nature, became simulacra of the significance of Jesus, both reinforcing the message of what that significance was and at the same time making it all but impossible to change the perception of it. Reading him through the lens of 'salvation from sin' became the accepted and commonly shared view of his life. And his death became a surrogate payment to an outraged God rather than a freely given gift to the life of the world.

This meant it became impossible, indeed unthinkable, to question publicly some of the implications of this view alluded to in previous chapters of this book. What kind of God, for example, would require the death on a cross of this good man to atone for the sins of others? Or, more specifically, could such a God be the good God Jesus related to, prayed to and spoke about throughout his life; witnessed to in his compassion for the suffering of others and revealed in parables about the nature of God's kingdom? A kingdom that is 'like a woman who takes yeast and hides it in three measures of meal ...' (Q 13: 20–1; Matthew 13: 33; Gospel of Thomas 96: 1–2).

The God of the creeds, in direct contrast to the God of Jesus, appears concerned only with the eternal salvation of an elect group of humans and unconcerned about the lives of all other beings. The God of the creeds, an 'almighty' Father and Creator, is theologically akin to the one who saved only Noah and his family and condemned all other creatures to live in fear and dread of their descendants. How can this God be called 'good'? How can we acclaim as 'the one, true God' one in whose name violence against our fellow human beings and all other beings has been and still is sanctioned?

Creeds as contexts for an imperial Christian mindset

But this God invoked in Church creeds – presumed to contain all we need to know about Jesus – is deeply imprinted on Christian consciousness. He has been, and is, made known to us through extended versions and theological summaries of the Pauline *kerygma* that instruct us in how we are to interpret not only the birth and death of Jesus but also the sacred texts of his own Jewish community. This method of teaching what in the beginning were largely illiterate communities was, and still is, used by clergy on particularly important religious occasions – such as baptism, sermons, funerals, seasonal liturgical celebrations and polemics against heretics. Over centuries, it has evolved into a systematic hierarchy of truths in regard to Jesus' birth and death.

This hierarchically structured model and its norms have formed part of the cognitive, symbolic and socialized Christian life-world: one in which participation in communal religious life and practice, based on accepted interpretations of the meaning of certain terms, has created a more-or-less agreed upon understanding of what constitutes legitimate questions and answers about Jesus. This consensus has so informed and sustained Christian faith and practice as to be a truly inconvenient, very real hindrance in formulating an adequate Christian response to the challenges of climate change. But it is more than inconvenient. It keeps Christian theological vision confined within the arid environment of an imperial mindset – one that deprived Jesus himself of life.

What do I mean by an 'imperial' mindset? At its simplest, it is a structural system of dominance over the world view of a particular community that works by controlling the content, meaning and modes in which common forms of knowledge are communicated. Over time, this creates a religious mindset in which certain truths about God are held to be self-evident and membership of the community depends on acceptance of those truths in a particular form.

This happens because the form our knowledge takes is environment-dependent, that is, it depends on relational processes between communal, social and physical bodies within a shared life-world. While knowledge is unique to an individual and no one can assume that everyone perceives phenomena in the same way, it is also a communal resource built up over time through networks of information and the latent emotional content that constitute the lived historical tradition of a community. Where that interactive relational element is present, symbolic behaviour and understanding of symbols, such as a crucifix, belong within a shared discourse in which imagination, thought, emotions, meanings and perception function coherently.

That discourse functions most constructively when it enables each person to generalize her spontaneous experience of the life-world. But on the basis of her inherent understanding and ability to interact with her environment, she must also be able to participate actively in the communal construction of knowledge by freely and responsibly denying or affirming its truth or relevance. Then interaction between interdependence with the environment of knowledge-building and the autonomy of the individual furthers the development and evolution of thought within a community.

In direct contrast, an imperial mindset presumes a relationship between power and knowledge where knowledge flows one way: from the powerful few at the top down to the masses at the bottom. The difference in power potential makes it possible for those at the summit to dominate individual experience and the articulation of knowledge of the world through vertical control of the relational structures of communication, decision-making and accepted values. Johan Galtung's work on nonviolence, in particular 'A structural theory of imperialism', has been particularly pertinent and helpful to me here. It was published originally in 1971 in *The Journal of Peace Research* and re-published since in many overviews of the structures of dominance, violence and dependency characterized by the vertical interaction patterns inherent in imperial relations.

Galtung defines imperialism as a relationship between a Centre nation and a Periphery nation typified by the *dominance* of the Centre. At the same time there is a *harmony of interest* between the centre in the Centre nation and the centre in the Periphery nation. There is more *disharmony of interest* within the Periphery nation than within the Centre nation and there is a *disharmony of interest* between the *periphery in the Centre nation* and the *periphery in the Periphery nation*. These characteristics of imperialism, in Galtung's view, mean that it is not merely an international relationship. Rather, it is a combination of *inter* and *intranational* relations (Galtung 1971: 83).

There are two basic mechanisms that sustain these relations. First, there is a vertical interaction between Centre and Periphery with the Centre occupying the leading position and the Periphery being subordinate and dependent. This is the major source of inequality. Second, the absence of any effective horizontal interaction between the Centre in the Periphery nation and its periphery is exploited, so as to leave those on the peripheries open to penetration, fragmentation and marginalization. It is worth stressing that this analysis of imperialism is aimed at exposing it as a *major form of structural violence*, whether the system being analysed is economic, social, political, epistemological or religious.

The Nicene Creed

Against this background we may look more closely at the historic context in which Christianity was taken over, in every sense, by imperial forces. In June CE 325 the Roman Empire had the role of Centre 'nation' and Christian bishops the role of Centre in Periphery 'nations.' This was the moment when Constantine summoned to Nicea the ecumenical council of bishops, that is, one inclusive of those from both the west and east of the Empire. His aim was:

> [T]o consolidate the Church, which represented in his eyes the spiritual aspect of his empire, on the basis of the widest possible measure of doctrinal unity.
>
> (Kelly 1950: 211)

The dominance of the Centre was typified in the person of the Emperor Constantine. He could and, by summoning the bishops, did presuppose a harmony of political interest between them and himself: with them functioning as 'Centres' within and, in some cases, on the periphery of the Empire. Therefore he determined to ensure a harmony of spiritual interest also by vertically unifying the doctrinal structure of the Church.

Before this council there was very evident *disharmony* of doctrinal interest among the bishops as they struggled individually with paganism, Gnosticism, Arianism and other disparate forms of Christian thinking in quite different community settings. The reason being that prior to this council, all creeds and summaries of faith were local in character and owed their immediate authority, no less than their individual stamp, to the liturgy of the local church in which they emerged (Kelly 1950: 205).

> Therefore Constantine's carefully phrased inaugural address – if we can trust the account given by Eusebius – fastened on the perils of internecine strife in the Church and voiced his ardent longing for peace and unity among the bishops. None of his audience could have been left in any doubt as to what was expected, and the part the emperor played in the day-to-day debates was of a piece with this ... While doubtless he looked to the council to manufacture a creed and was not behindhand in giving a lead to the drafting committee ... what made it momentous in import was the ecumenical character of the gathering and the altogether new position of the Church *vis-à-vis* the state.
>
> (Kelly 1950: 211–12)

Creeds as mechanisms of structural imperialism

Several pertinent points emerge from this necessarily brief account of what became known as the Ecumenical Council of Nicea, one that included Churches from both eastern and western regions of the Roman Empire.

First, the doctrinal Centre of Christianity became effectively synonymous with the political Centre of the Empire. The emperor persuaded three hundred and eighteen bishops, representing churches from both the eastern and western reaches of that empire, to accept a formula of faith. That became known as the Nicene Creed. It could equally well have been titled the Imperial Creed: an *inter* and *intra* national statement of imperial spiritual belief imposing harmony through doctrinal unity. This vertical interactive pattern meant that adherence to the creed could be used as a criterion for the admission of clergy and their congregations to, or their exclusion from, 'the new state Church', just as an oath of allegiance to the State served and still serves in many countries as a criterion for citizenship.

By the late fifth century CE the two forms of allegiance often merged into one, especially in western parts of the Roman Empire carved up among barbarian invaders. When Clovis, leader of the Franks, was converted to Christianity and baptized, so too were his people in the nation that expanded into what is now France. By the eighth century, when Charlemagne was ruler of France and his territories extended to parts of northern Italy, northern Spain and what is now west Germany, their inhabitants too were converted (often forcibly) to Christianity.

> Failure to be baptised meant disenfranchisement and virtual outlawry. Citizenship and being a Christian were synonymous.
>
> (Nineham 1993: 7–10, 28)

Second, the Creed served a religious function that reinforced the dominance of the bishops, as well as that of the Emperor, over the lives of the Empire's citizens. This ensured a harmony of interest between those at the Centre and those on the periphery of the Empire. The very practical outcome from this was that the organization of the Christian Church followed Constantine's reorganization of the Empire into twelve dioceses, of which the largest, Oriens, the praetorian prefecture of the East, embraced sixteen provinces and the smallest, Britain, just four. The international and intranational systems of communication between bishops' dioceses were modelled on those in the Roman civil service and areas of ecclesiastical administration coincided in name with those of the civil administration.

The imperial spiritual power of the Church over the life and death of its subjects rested on its claim to administer eternal salvation: that is, to judge and to punish those members who did not conform to Church laws by excluding them from the salvation from sin offered in church sacraments – and so from eternal life. The chosen instrument for imposing that power was adherence to, or departure from, the truth of the creedal formulas. As the Western Christian Church expanded and established its power base in Rome so did its power of exclusion. It anathematized, that is, excommunicated certain classes of people within its own religious boundaries as well as vast numbers belonging to 'peripheral' religions: those classed as pagans, idolaters, magicians, witches, heretics, Jews, Muslims and later, as we shall see, other Christians.

Simone Weil sombrely remarked on the continuing power of this imperial legacy:

> After the fall of the Roman Empire, which had been totalitarian, it was the Church which was the first to establish a rough sort of totalitarianism in Europe in the thirteenth century, after the war with the (heretic) Albigenses. This tree bore much fruit.
>
> And the motive power of this totalitarianism was the use of those two little words: *anathema sit*.
>
> (Weil 1951: 48)

The third notable structural mechanism that developed out of these imperial interactions was a Christian potential for violence of all kinds: military, physical, legal, territorial and spiritual. The military character of the ecumenical gathering at Nicea and the altogether new position of the Church vis-à-vis the State is succinctly described by John Dominic Crossan. He quotes Eusebius describing the Imperial banquet held to celebrate the conclusion of the Council:

> Detachments of the body-guard and troops surrounded the entrance of the palace with drawn swords, and through the midst of them the men of God (the bishops) proceeded without fear into the innermost of the Imperial apartments, in which some were the emperor's companions at table while others reclined on couches arranged on either side. One might have thought that a picture of Christ's kingdom was thus shadowed forth.

Who, asks Crossan, other than an imperialist Christian, could have thought that this was a picture of Christ's kingdom, of the Messianic banquet foreshadowed in Jesus' table fellowship with the poor, with women, with the outcast and the sinner? (Crossan 1994: 201).

The potential for Christian violence implicit in this picture became, in the centuries following that ecumenical Council, the mechanism that destroyed the political and spiritual unity that was that Council's own stated aim and, for a time, its achievement. Constantine himself began the process of break-up by founding a second imperial capital in the East, Byzantium, alongside Old Rome in Italy. The barbarian invasions into the Latin West destroyed its political and linguistic unity with the Greek East; a severance carried a stage further by the rise of Islam in the Mediterranean and by Charlemagne setting up a 'Roman' Empire of his own. He was crowned Emperor by the Pope on Christmas Day in the year 800.

Refused recognition by the Byzantine Emperor, Charlemagne retaliated with a charge of heresy against the Byzantine Church, denouncing the Greeks for not using the Latin term *filioque* in the Nicene Creed. Not part of the original Greek text, this was the term used in the West to describe the 'procession' of the Holy Spirit from the Father and the Son – as it still is. Behind this denunciation lay a Roman Papal claim to spiritual power over the whole Church, East and West. As soon as the Pope tried to enforce this claim within the Eastern Patriarchates, trouble was bound to arise. Adding the *filioque,* or any other clause, to the Creed – for so the Greeks saw it – was specifically forbidden by the Ecumenical Council of Nicea. Nothing short of another ecumenical council could make such a change. And by now, linguistic as well as theological and cultural barriers made such a meeting all but impossible.

So, too, did they make it harder to maintain the unity of the Church's religious structures. By CE 863 there was an open breach between the Churches of Rome and Constantinople that, by 1054, had become 'The Great Schism'. Even then, there were hopes on both sides that the misunderstandings between them could be cleared up. A final attempt to do so, when Pope Leo IX sent three legates to Constantinople, ended with them laying a Bull of Excommunication against the Patriarch, Michael Cerularius, on the altar of the Church of the Holy Wisdom there. Yet ordinary Christians in the East and West remained largely unaware of the dispute and its esoteric theological origins (Ware 1984: 67).

The lasting effects of credal imperialism

However, this ignorance of such disputes and their consequences was not to last:

> It was the Crusades which made the schism definitive: they introduced a new spirit of hatred and bitterness, and they brought the whole issue down to the popular level.
>
> (Ware 1984: 67)

The Crusades or 'Wars of the Cross' ushered in a new era of Christian imperial, militarist violence whose reverberations resound to this day. Although usually understood as being directed against Islam, one of the most shocking incidents in their history occurred in 1204 when the Western Crusaders, originally bound for Egypt, were persuaded by Alexius, son of the dispossessed Emperor of Byzantium, to turn aside to Constantinople in order to restore him to the throne. Eventually the Crusaders lost patience at the vacillations of the imperial party and sacked the city in three days of pillage never forgotten by Eastern Christendom.

> 'Even the Saracens are merciful and kind', protested Nicetas Choniates, 'compared with these men who bear the Cross of Christ on their shoulders.' How could men who had specially dedicated themselves to God's service treat the things of God in such a way?
>
> (Ware 1984: 68–9)

This all-too brief account of a very complex historical situation, its development and religious consequences does, however, give some idea of the role played by the Nicene Creed in structuring Christian Church life along imperial lines. It also gives some indication of the way in which the struggle for spiritual power, when allied with the power of the State, justified and continues to justify the most atrocious violence against others: whether they are Christians, or belonging to other religions, or to none.

Theologically, and indeed visually, this growing strand of violence can be traced throughout the Western Christian iconography of salvation over the past two millennia, centred on Jesus as a militant and victorious saviour figure. One of the earliest and simplest symbols, IHS, has been used continuously in church architecture, on altar adornments and on clerical vestments. Christians have been taught that this represents the name of Jesus, in an abbreviated form of the Greek version of his name, which itself derives from the Hebrew *Jehoshua*, meaning 'Yahweh [God] saves'. The use of this Latin acrostic became widespread throughout Western Europe where it was variously interpreted as *Iesus Hominum Salvator* (Jesus the Saviour of Mankind), or *In Hoc Signo* [*Vinces*] (in this sign you shall conquer).

This latter interpretation goes back to the myth of Constantine's infamous dream in which, he said, an angel appeared to him bearing a military emblem in an abbreviated form (*chi, ro*) of *xristos*, the Greek for Christ. Constantine had this emblem stamped on his coins and woven into the banners of the imperial army. It was used in its Latin form during the Crusades. This mutation of the name 'Saviour' into a militarist, triumphalist emblem, and of the nature of salvation as a militarist form of victory over sin and death, has a long and violent imperial religious history that, alas, continues to this day.

It chronicles and reveals a lethal inversion of Jesus' command to love one's enemies into a licence to kill those Christians *defined* by other Christians as enemies: either by virtue of their birth, education or religious conviction. This intra-Christian history has formed, or rather deformed that of my native Ireland right up to the present day. And we are now seeing such religiously legitimized violence play out on a world stage against those who have never heard any formulation or definition of Jesus' origins and death.

Looking as steadily as possible at this history is always a grim experience. In a time of climate change it reveals two major issues to be addressed by Christian theologians. The first arises by omission: nowhere in that history is any attention paid to or concern expressed about the effects of human violence on the environment as presently understood. The second question arises by implication. Is there any way in which present work on the historical Jesus offers resources for what Crossan calls the process of reconstruction: something that must be done over and over again in different times and different places, by different groups and different communities, and by every generation again and again and again (Crossan 1999: 306).

11 What Jesus Said

The problem the historical Jesus poses is really not that we do not know what he had to say, but rather that, when we get wind of it, we do not know how to handle it. Rather bear those ills we have than fly to others that we know not of! And so the church prefers to whistle in the dark.

(Robinson 2007: 226)

When we find that the Jesus of history is really not the imperial, militantly apocalyptic saviour of an elect few but an historical figure whose sayings have been recorded, then, says Robinson, we are left not knowing how to handle what he said. This conclusion emerged as demonstrably true in the preceding chapter's summary of the history of imperial Christianity. It stands as record of a conspicuous moral failure to handle what we know Jesus said about how we are to relate to those who regard us or whom we regard as enemies:

> Love your enemies and pray for those persecuting you, so that you may become children of your Father; for he causes the sun to rise on the bad and the good and rain to fall on the just and unjust. And if you love those who love you, why should you be commended for that? Even the tax collectors do as much, don't they? And if you greet only your friends, what have you done that is exceptional? Even the pagans do as much, don't they? To sum up, you are to be as unstinting in your generosity as your heavenly Father's generosity is unstinting.
>
> (Q 6: 27–8; 35c–d; 32; Matthew 5: 47–8; Luke 6: 27–8, 32–5)
> (Funk 1993; Robinson *et al.* 2002: 83–5)

Divine impartiality

To the extent that we can recover Jesus' sayings from the tradition, these are close to being at the heart of them. In particular, scholars have ranked the command to love one's enemies third highest among sayings that almost certainly originated with Jesus (the other two include the complex with a similar message, about turning the other cheek, Matthew 5: 39–42; and the cluster of beatitudes, Q 6: 22–3; Luke 6: 20–2). One of the reasons for their decision is that the injunction to love enemies is a *memorable* aphorism, remembered precisely because it cuts against the social grain and constitutes a paradoxical truth: that those who love their enemies have no enemies (Funk 1993: 145–7).

So how have those of us who call ourselves Christians handled this saying? Historically, by ignoring it; or more reactively and more consistently by personal and institutional denial of its entire intent. And by then claiming the right to act in direct opposition to it in Jesus' name. But when we act like this, Jesus tells us, we are refusing to behave as recognizably children of his Father. For we are commanded to love our enemies precisely because this Father does *not* discriminate, as we do, between those we consider worthy of love and those we consider worthy of hate. All such distinctions vanish before One who causes the sun to rise on both the good and the bad and the rain to fall on just and unjust alike. And, we must remind ourselves, neither does that Father discriminate between the human and other-than-human recipients of sunshine or rain.

The fact that the cosmic background of life is apparently neutral and therefore inclusive serves as the basis of a fairly radical notion of God. A God who treats all human beings even-handedly is not much in evidence in the Bible. Indeed the God pictured there is, for the most part, highly partial and often quite vindictive. But that is not to set up in any way a binary opposition between the spirit of love, as evidenced in Jesus, and the dead letter of the law as belonging to Judaism. On the contrary, the radicalizing tendencies within the preaching of Jesus represent a reform of Judaism by Judaism, a self-correction interior to Judaism that drew upon its deepest tendencies. There is nothing more Jewish than what Jesus said about God (Caputo 1997: 226).

His insistence that God shows no partiality (Deuteronomy 10: 17) formed part of a vision that survived into the first Christian communities and was a crucial factor in the admission of Gentiles to baptism (Acts 10: 34f.). In a world emerging from tribal cultures, this practical implementation of Jesus' admonition to love enemies must have struck his hearers as truly radical. No longer was ethnic identity to be the primary focus. Instead, what human beings had in common (the operation of the elements), rather than what defined their differences, became the central criterion for our relationships (Funk 2002: 59–61).

Therefore, the impartial nature of God's love for the world is acted on in Acts 10: 34, stated in Romans 2: 11 and used as an argument for opting for the poor in James 2: 1. The Church however, in its imperially militant and apocalyptic modes and mechanisms, has systematically gone against this vision by decreeing that one is only identified as a child of God through baptism and its accompanying pledge of adherence to certain doctrinal formulas. Those who do not adhere to them are not considered to be God's children and are, therefore, all too often seen by the Church as legitimate enemies.

Since the reign of Constantine the consequences of implementing this discriminatory policy are recorded in the history of internal battles between Christians; between orthodox Christians and those they designate as heretical enemies such as Docetists, Arians, Albigensians and Waldensians; then in intra-Christian battles before and after the Protestant Reformation and so on to the present day. In 1967 the Presbyterian Church in New Zealand charged Professor Lloyd Geering with heresy and in 2002 Dean Andrew Furlong was tried for heresy by the Church of Ireland (Furlong 2003). Geering's case was dismissed and Furlong resigned after agreeing to an undisclosed settlement. Their outcomes aside, the fact that these trials occurred at all tells us much about official contemporary Church policies, as do earlier conflicts between Christians that have such enduring and visible consequences as steel and concrete barriers between Protestant and Catholic communities in Northern Ireland.

For the injunction to love enemies not only cuts against the social grain: it cuts more incisively still against the religious grain of an imperial apocalyptic Christianity. It seems impossible for us to accept or to abide by Jesus' saying that God does not distinguish between the just and the unjust and sends the rain and sunshine necessary to sustain life indiscriminately on both. In other words, to accept that God does not restrict divine love to those that the Church, on its own criteria, judges to be morally superior to others and so worthy of enjoying the reward of their labours in heaven. By default, those who fall short of ecclesiastical standards are at best left to God's mercy and at worst, pronounced condemned to hell and eternal punishment.

A truly shocking contemporary example of this occurred within the past two years at the funeral of a Roman Catholic priest working in Latin America, who, against official Church policy, had allowed indigenous religious rituals to be integrated into the celebration of the Eucharist. Officiating at the funeral, his bishop declared that while he did not know whether or not the priest was in Purgatory or in Hell, he was quite certain that he was not in Heaven. We can be quite certain that such (post)colonial attitudes are against the grain of what Jesus said.

Further distressing examples of an arrogated right to condemn others to hell as 'enemies' of God can be found in ecclesiastical tirades against Jews, Muslims, practitioners of paganism, animists, those from indigenous religions, homosexuals, lesbians, and others. In many cases, they understandably respond by expressing their hatred of Church members.

Such ecclesiastically formulated expressions of hatred may appear to be the reverse side of the coin of love. But instead it demonstrates the paradox that, according to Jesus, that currency has no reverse side. Loving those that hate you means that you no longer consider them enemies. The force of this paradox, and its cross-grained social and religious effect, appeared very clearly when, in 2007, an English Quaker peace activist, Norman Kember, was captured by an Islamist group, paraded blindfold on television and threatened with death. After his unexpected release his refusal to use the expected rhetoric of hate and retribution about his captors confounded his audience.

But why, asks Jesus, should you be commended for loving those that return love to you? That is no more than an economic exchange: like those between tax collectors. If you greet only your friends, what's exceptional about that? Even those we call pagans do as much. Rather, love those who act hostilely or appear inimical towards you and so be identified as children of the Father. *Love your ene-mies* is probably the most radical thing Jesus ever said. Unless, of course, one considers the parable of the Samaritan. There the admonition is to let your ene-mies love you (Funk 2000: 16).

Yet in the preceding chapter we saw how religious structures have been put in place to do precisely the opposite: to identify others as enemies so as to legitimize an inimical response to them from us. Formally expressed in witch burnings, bulls of excommunication, heresy trials, exile, torture and death, this sanctioned hatred has been expressed and executed in the name of the very God whose indiscriminate love for all is the criterion set by Jesus for our relationships. Our enemies cannot, indeed must not, be classified as his Father's enemies. For we only become his children by loving our enemies and, by doing so, no longer cat-egorizing them as such. The God of Jesus is one who has dispossessed himself of the power to hate any of his creatures.

Some theological implications of loving our enemies

How do we begin to handle this most inconvenient truth? At the very least, by refusing to support an imperial theological mindset whose discriminatory images of our fellow beings are directly opposed to the truth of what Jesus says about them. How can we support it in a time of climate change when we know, indeed are directly experiencing, some of the ills that are the legacy of the structured violence it endorses? Those ills have fallen, as we know, not only on us and on our potentially blameworthy fellow humans, but on blameless communities of other-than-human living beings and their environments. Their all too under-standable 'fear and dread' response to us follows, for the most part, from our treating them as 'brute beasts' incapable of judging right from wrong. So we claim moral superiority over them in order to justify treating them amorally.

Jesus, however, confronts us with the absolute demands of an ecology of love that does not discriminate between us on the basis of our conduct; and certainly does not do so by discriminating between us and those who conduct themselves differently from us. Those fellow creatures who differ from us by virtue of their differing evolutionary histories and are coupled with us in complex human and natural systems, do not appear to, indeed cannot discriminate between us on any other basis than that of how we treat them.

Catherine Keller draws attention to the way in which the postcolonial theory of Gayatri Spivak, in its specific preoccupation with an older imperial history, reaches out against the grain of this world and envisions one that is 'ecologically just'. It exists through a different kind of unification from that of a militarily imposed one. This ecological unification, I would argue, is already there in our belonging to the larger community of life on earth, although our conscious engagement with it is, as Spivak discerns, conspicuously lacking.

Keller herself discerns a metamorphosis of Christianity also taking place in this postcolonial space. It is a metamorphosis into a form not interested in religious triumph but in the 'peace, justice and integrity of creation' named in the World

Council of Churches programme of the 1990s. This form of Christianity recognizes that 'we are all already bound up in our earthly ecology.' She quotes Thomas Mann's saying that 'love cannot be disembodied even in its most sanctified forms, nor is it without sanctity even at its most fleshly ... Love is our sympathy with organic life, the touchingly lustful embrace of what is destined to decay.'

So in a conclusion that harks back to what has been said about the unconditional love at the heart of the kingdom of the God of Jesus, she insists that there would be nothing *theologically* accidental about the emergence of an ecological motif precisely there where love begins to overflow within an infinite ecology of relations.

> Once we free theology from a disembodied paternity, once, in other words, it discovers the open-ended creativity in which all creatures partake, it can disclose the widest context of our embodied life. That creaturely width exceeds the scale of any empire, and yet in the history of the west it has been bound and defined by the succession of imperial powers. Nevertheless, Christianity has always addressed a spacetime beyond any particular context ... In other words, this universal vision has lent itself both to empire and to anti-imperialism ...
>
> (Keller 2005: 126–33)

What is being envisioned here, in a time of human-induced climate change, is a theological attempt to try to think through the death of a common perception of God: a God of imperial absolute power. Caputo characterizes this attempt as a postmodern theology of the Cross: one that does not think in terms of some celestial transaction going on where there is a settling of accounts between divinity and humanity, as if this death is the amortization of a debt of long standing and staggering dimensions.

> If anything, no debt is lifted from us but a responsibility imposed upon us. For we are laid claim to by this spectacle, by the cry against unjust persecution that issues from the dangerous memory of this scene, by the astonishing spectacle of greeting hatred with love, of answering persecution not with retribution but with forgiveness ... Insofar as there is any philosophical life left in this increasingly dated expression, the death of God, it refers to an ongoing and never finished project of deconstructing the God of ontotheo-logic, which is for me above all the God of sovereign power.
>
> (Caputo 2007: 66–7)

What is happening here is a shift in theological focus from what we say Jesus did *for* us – 'dying for our sins in accordance with the Scriptures' – to what Jesus said *to* us: 'love your enemies.'

The unconditional gift

Nowhere does the unconditionality of gift within the kingdom of gifts strike as hard against the grain as in the matter of forgiveness of those we perceive as enemies. Ordinarily, we expect bad behaviour towards us to be punished, sometimes severely, and good behaviour rewarded. Forgiving, however, means giving away what is due to come back to us, whether it is a debt or an obligation incurred. Jesus, however, in this, as in other matters, reverses expectations and makes forgiveness reciprocal. 'Forgive and you'll be forgiven' (Luke 6: 37). One cannot be a recipient of forgiveness unless or until one is an agent of forgiveness. It's as simple and as difficult as that (Funk 1993: 3–10).

Just how simple and how difficult it is forms the basis for Derrida's reflections on the workings of the post-apartheid Truth and Reconciliation Commission in South Africa. Behind that historic event, of course, stands the aftermath and effects of other twentieth-century genocidal atrocities, most notably the European Holocaust. Here it becomes clear that forgiveness (given by God, or inspired by divine prescription) must be, says Derrida, a gracious gift, without exchange and without condition. Sometimes, however, it requires as its minimal condition the repentance and transformation of the sinner.

> What consequence results from this tension? At least this, which does not simplify things: if our idea of forgiveness falls into ruin as soon as it is deprived of its pole of absolute reference, namely, its unconditional purity, it remains nonetheless inseparable from what is heterogeneous to it, namely the order of conditions, repentance, transformation, as many things as allow it to inscribe itself in history, law, politics, existence itself. These two poles, *the unconditional and the conditional* are absolutely heterogeneous, and must remain irreducible to one another.
>
> (Derrida 2001: 44)

Behind this statement lies the insight, repeated here so often in as many ways as possible, that the pure gift, the gift pure and simple, is im/possible, conditioned by possibility *and* impossibility. That is why giving is always for-giving (Caputo 1997: 226). Pure and unconditional forgiveness, in order to have its own meaning, must have no meaning, no finality, even no intelligibility. It is a madness of the impossible (Derrida 2001: 45)

Yet how are we to live with the terrible effects of the historic events considered here if not by aiming for this giving that is always for-giving? How are we to respond to what Jesus said about loving our enemies if we do not learn how to forgive? How do we move beyond the punitive vision of a God who wreaks vengeance on his enemies to one who forgives us as indeed we forgive others?

The hinge of this famous saying is 'as indeed', *hos kai* (Matthew 6: 12), forgive us as indeed we forgive; dismiss our debts as we dismiss our debtors. Dismiss our creditors as indeed we give away our credits.

> We give our credit away absolutely, unconditionally, without the expectation of return. God supplies the rest, the supplement we dare not desire, which can be granted only if we do not desire it, only if we put it out of our minds. The yield of giving is more giving. Giving gives giving. God is the name of the giving that spreads like a fire, or that runs like water over the land, that multiplies the loaves of giving to infinity: so forgiving breeds forgiveness and breaks the circle, the cycle of vengeance, and, beyond vengeance, of simple debt.
>
> (Caputo 1997: 226–7)

I agree with Keller and Caputo that deconstructing the God of sovereign power is an ongoing and never finished project. I also agree with Crossan's view, cited at the end of the preceding chapter and taken for granted by Robinson, that such a deconstruction requires us to engage in reconstructing the past of the historical Jesus and the earliest records of what he said. Otherwise we are left 'whistling in the dark' – not a good place to be when trying to envision change!

Such scholarly historical reconstruction can be seen as a way of enabling us, as well as requiring us, to 'handle' what Jesus said as constructively as possible. For just as postcolonial theory and postmodern philosophy are supplying theological resources for dealing with religious imperialism and its enduring effects, so too – as I indicate in these chapters – is biblical scholarship. And the fruits of this scholarship is one of the contemporary effects of the historic event of the Reformation for which we must be truly thankful. Whereas until modern times it was assumed that, on the whole, people could only know about Jesus through their experience of Church life and, in particular, through Church creeds, Robinson and other scholars are confident that today we have the means necessary for this reconstruction: that we can also know what Jesus had to say. The givenness of Jesus has been given back to us.

This underlies Robinson's conviction that, if we so wish, we can discover what Jesus had to say. The difficulty we have in handling its anti-imperial character makes it imperative that we do so.

Excavating Jesus' sayings

Crossan, Kloppenborg and others use the analogy of 'excavating Jesus' to describe the digging down exegetically amidst the texts now available to us in order to reconstruct his life and the world in which he lived – and so place his vision and what he said in context. Compared, however, with actual excavations where, even if you disagree with the chosen list of sites, you must agree that they all exist and can be seen somewhere on the ground or in a museum, the same is not true for some of the items on the exegetical list (Crossan and Reed 2001: 6–14).

Nevertheless, just as methods of correct layering and accurate typology of remains are all-important in site excavation, so too the results of stratigraphic analysis of relatively recently discovered texts (such as those of Jesus' sayings found at Nag Hammadi in Egypt in 1945) and of critical theory applied to them can be defended. Form criticism establishes the earliest formats – a parable, an aphorism, a dialogue, a law etc.– used in transmitting the tradition orally. Source criticism establishes who is copying from whom. Redaction criticism builds on such copying to establish the purpose of the copyists' omissions, additions or alterations. Tradition criticism uses all of the above to establish the successive layers of the tradition's development. But it is probably source criticism, above all else, that underpins the problems and at the same time highlights the importance of exegetical layering.

> For example, if Matthew and Luke creatively copy Mark, and if John very, very creatively copies those three earlier texts, what follows? Instead, for example, of concluding that Jesus' entry into Jerusalem the week before his death is told in all four gospels (independently), we must conclude that it is told in three super-imposed layers all based on Mark (dependently). That immediately raises another question. What historical layer is Mark's account? Is it history from a layer dated to the late 20s or parable from a layer dated to the early 70s CE?
>
> (Crossan and Reed 2001: 12–14)

Attempts to answer such questions centre on the typology of textual remains through which the 'excavating' of the historical Jesus takes place. The basic type or layer of these is what Crossan and other scholars call a 'sayings gospel': one composed primarily of words attributed to Jesus. These include aphorisms, parables and short dialogues. Incidents, insofar as they are present, emphasize the word rather than the deed. There are few miracle stories, no passion narratives and no post-resurrection apparitions in sayings gospels. The classic examples from the middle of the first century CE are the reconstructed 'Q' Gospel and the Gospel of Thomas. The latter exists in textual form in a fourth-century copy translated from Greek into Coptic (Crossan 1999: 294).

The sources given earlier in this chapter for the biblical sayings about love of enemies assume that the layer in which they occur was copied creatively in different settings by the authors of the Gospels of Matthew and Luke. 'Q.' (with a full stop making it clear that it was meant as an abbreviation of *Quelle*, 'source') was first used for this layer in 1880. By the 1890s it came to be used simply as a symbol for a source used by Matthew and Luke. Now that shared source has been reconstructed as a critical text resulting from the work, since 1985, of the International Q Project. The reconstruction, together with an informative history of the Project, is available as *The Sayings Gospel Q in Greek and English* (Robinson *et al.* 2002: 23).

The fruits of this scholarship, however, have not, for the most part, influenced either the lives of Christians or their understanding of Jesus' life and death. Nor has the fact that the authors of the four canonical Gospels, when using Q and other sources, had quite different viewpoints and purposes from those of Christians today when they altered, omitted or added to them in their accounts of Jesus' death. Instead the Pauline version, on which is based the kind of imperial theology of the Cross described and rejected here, continues to present us with a God of sovereign power engaged in settling accounts with humanity through the death of an innocent man.

So what difference would it make to reconstruct the sayings of the historical Jesus and use these as the basis for Christian theology in a time of climate change? Could these scholarly resources function as a postcolonial, mobilizing discourse of love that, according to Spivak, would have a worldwide remit comparable to that of liberation theology? She is in no doubt that this discourse can be learnt from the practical ecological philosophies of the world. What deserves the name of love, she says, is the necessary collective effort to change laws, relations of production, systems of education and health care (Keller 2005: 131).

The practical ecological component in this collective effort is increasingly evident in shared scientific studies of our complex interactions with the other-than-human members of the community of life on earth. A recent landmark study, 'Complexity of Coupled Human and Natural Systems', integrates the work of ecological and social scientists internationally. They are conscious that their human co-operation simply mirrors the fact that, as the study shows, as globalization intensifies there are more interactions among even geographically distant systems. Therefore, their report concludes, it is critical to move beyond existing approaches for studying coupled systems (Liu, Dietz *et al.* 2007: 1513–16).

Love and liberation theology

Implicit in that conclusion is the fact that the ultimate human-natural coupling on a global scale emerges as what we call climate. This is the planetary unifying factor that binds all our lives and our fates together. It is also true, however, that the discourse used in the above report to describe the impacts and effects of human-nature couplings is not supplemented with that of love. Nor, given its origins, could one reasonably expect that to be the case. But that does not mean that its consequences and implementation are not important for how we interact with each other here and now as we shape new patterns of land-cover change, human population distribution and human activities. For a key area of concern is the variation in legacy effects, from decades to centuries, of interactions between human and natural systems.

To talk of legacy effects is another way of saying that how we live today affects all those who will live after us: just as we are living with the legacy effects of the historic events detailed in earlier chapters. However Spivak's other, rather unexpected appeal to a liberation theological model to bring about an ecologically just society brings these rather abstract notions down to earth. At the same time it connects explicitly with a 1991 essay by Robinson entitled 'The Jesus of Q as a Liberation Theologian'.

While acknowledging that the title is as problematic as trying to recover genuine mosaic stones and then fit them into the same coherent picture from which they originally came, there is, he said, an overriding *directionality* to the earliest stages of the Jesus tradition that, when analysed in reversed chronological sequence, points towards the point of origin: the approximate position where

Jesus would have been located. Back to Jesus means going this way, not that way. In my terms, towards peace, not violence. This points to his relative position within Judaism: to the fact that he was more Jewish than the ensuing tradition; less Christian than, for example, the canonical Gospels. Thus the way he points, as a role model, as an authority, as a precedent, may be clearer than the actual biographical facts. Hence Jesus' relevance for us may not ultimately be dissolved by the obscurity in which his biography lies.

Indeed, warns Robinson again, the direction he points may be ascertained with sufficient clarity to be uncomfortable. Here it directs us back to important metaphors in Q that deal with such mundane matters as bread and stones. Catching sight of these metaphors, one should visualize a small loaf of bread and a fist-size stone, familiar to Palestinians from the time of David and Goliath through to the intifada. The Jesus of Q thought of loaves roughly resembling stones, much as a fish resembles a snake (Q 11). His trust in God was to the effect that God, like a human father, would not replace the daily loaf for which one prayed with an inedible stone. The painful human point of departure in the Q Prayer (a precursor of what is known as the canonical Lord's Prayer) commonly has at its centre a loaf of bread: 'Our day's bread give us today' (Q 11: 12; Matthew 6: 10).

One may, says Robinson, recognize the petition 'thy will be done' as a secondary Matthean interpretation, a moralizing addition to 'let your reign come'. Whereas in Q the petition for God's reign to come precedes that for the loaf of bread. It makes God's reigning as specific as a day's ration. The petition for bread is meant literally. The Q people had been led to expect that trust in the coming of the kingdom would involve daily bread. The emphasis is on sustenance as a gift of God, yet the loaves are not spiritualized away in Johannine style, as manna from heaven (Robinson 2005: 451–4).

Stepping back from this, so to speak, Robinson's vision of what he calls 'the Q people', the ones who remembered what Jesus said and what it meant to them when he said it, chimes in with Galtung's analysis of those in our own time who live on the 'periphery' of capitalist empires. These are the people who, living 'on the margins', have aroused the passion and concern of liberation theologians fired by a vision of what 'God's reign' means for the poor. In his introduction to Franz Hinkelammert's description of North-South political economics, *The Ideological Weapons of Death: A Theological Critique of Capitalism,* Cornel West says that academic theology in the economic North remains preoccupied with doctrinal precision and epistemological pretension. Yet for those Christians deeply enmeshed in and united with the poor in the economic South, 'theology is first and foremost concerned with urgent issues of life and death, especially the circumstances that dictate who lives and who dies'(Hinkelammert 1986: v).

Orthodoxy's consistent defence against the critique of liberation theologians has been its characterization of liberation theology as Marxist. Dom Helder Camara recognized this strategy: 'When I give food to the poor they call me a saint. When I ask why the poor have no food they call me a communist.' This theologic was the product of a Christian Cold War ideology, where the 'capitalist empire interpreted itself as the Christian (Western) world, a reign of God facing a reign of atheist evil'. So in his *Instruction* of 1984, the then Cardinal Ratzinger said that liberation theologies lead to 'a disastrous confusion between the *poor* of the Gospel and the *proletariat* of Marx'. But, asked Juan Luis Segundo in reply, if the poor are victims of oppression, why would the 'proletariat' not belong to this category? Only because Marx defended them and therefore, very improperly, they may be called the 'proletariat of Marx'? (Segundo 1985: 181).

The anti-imperialism of liberation theology restored importance to those on the peripheries; to a neglected level of existence (neglected that is, in traditional theologies). It used Jesus' singling out of the poor whose primary concern was how to find their daily ration of bread, not only as a unifying principle for their theology but also for their world view. Sin, according to Sobrino, reveals itself in historical form through the death that human beings inflict on one another. 'Unjust structures bring death near and inflict it daily.' Both these factors – the destruction of life for the masses and structural injustice as its cause – are central to understanding liberation theologians' 'option for the poor'. Neutrality is not an option (Segundo 1985: 117; Sobrino 1985: 164–9).

In an essay entitled 'Option for the Poor' Gustavo Gutierrez brings us back to Robinson's challenge and to the subject of this chapter by pointing out that it is not 'optional' in a theological sense either, since we owe love to every human being without exception. Nor does a commitment to the poor and oppressed rest ultimately, for Christians, on any social analysis, on our compassion or on the direct experience we may have of poverty. All these are valid reasons and play an important role in our commitment and, I would add, in being able to explain it to ourselves and to others. But for Gutierrez, as for all who hear what Jesus said, that commitment is ultimately based on the paradoxical nature of God's love. We opt for the poor not because they are morally or religiously better than others but because God makes no such distinctions (Gutierrez 1996: 27–34).

An unexpected and authoritative insight into Robinson's vision of the Q people comes from Eric Auerbach, one of the great literary scholars of the twentieth century. In a major critical work, *Mimesis: the Representation of Reality in Western Literature,* his criterion for selection was representation of the reality of the lives of the common people. Early on in his text he draws a comparison between the literature of antiquity, as represented by the approximately contemporaneous work of Petronius and Tacitus, and the version of Jesus' arrest and Peter's denial in Mark's Gospel. Recounting Peter's meeting and conversations with a servant girl and some bystanders, he remarks that:

> The incident, entirely realistic in regard to locale and *dramatis personae* – note particularly their low social station – is replete with problem and tragedy.

As a fisherman from Galilee, of humble background and the humblest education, Peter's meeting with the others is nothing but a provincial incident, an insignificant local occurrence. Viewed superficially, it is a police action that takes place entirely among everyday men and women of the common people. It portrays something neither the poets nor historians of antiquity ever set out to portray: the birth of a spiritual movement in the depths of the common people, from within the everyday occurrences of contemporary life. They thus assume an importance they could never have assumed in antique literature (Auerbach 1953: 41–7).

When the Gospels – whether canonical or sayings – set Jesus' life within those of ordinary men and women, their lives assumed an importance they could never otherwise have had. Moreover, what Jesus says to them and about them also assumes added importance. For he speaks not as an observer of their lives but as one who shares their 'low social station'; as someone on the periphery and not at the centre of power. As one of them, he experiences the everyday anxieties of finding enough food and water to live, of paying unjust taxes, of being ill and not cared for; of being at the mercy of imperial economic and military powers.

In a time of climate change when, as scientists predict, extreme weather events become more frequent and their effects more severe, and if, as predicted, Africa and Asia bear the brunt of climate impacts along with small islands and Asian river mega-deltas, the poorest peoples on earth will, like Jesus and the Q people, struggle daily to sustain life. Taking heed of what he said, their plight must affect the directionality of our theology. As it does we become more and more conscious that because we consume more food than we need, our legacy to them will be hunger; because we have hoarded raw materials and wasted energy, we shall leave them cold, ill and naked; because we have profited from shares in bottled water companies, they will die of thirst (Matthew 25: 31–46).

Or we can consciously reverse this trend by entering positively into the mystery of giving and, without knowing who the ultimate receivers may be, try to provide the conditions that make life-giving and receiving possible for future generations. Then the legacy of life-enhancing threads that attach us to all those who came before us remains unbroken and connected to all those who come after us.

12 Beginning Something New

The discoverer of the role of forgiveness in the realm of human affairs was Jesus of Nazareth. The fact that he made this discovery in a religious context and articulated it in religious language is no reason to take it any less seriously in a strictly secular sense ... Only through this constant mutual release from what they do can men remain free agents, only by constant willingness to change their minds and start again can they be trusted with so great a power as that to begin something new.

(Arendt 1958: 239–40)

Throughout this book some historic events and their effects have been discussed in the context of the present event of climate change. This is a new context for human history, a contemporary event with a global compass whose effects now call for change in us too. The change required is from a self-image dominated by a perception of ourselves as having the *right* to use earth's resources independently of the needs of all other beings. That in turn requires theological change: from an image of God as granting us the power to do this. Together, these images have contributed to a theological climate that in turn has sanctioned activities now shown to be dangerously close to destroying the planetary life support systems on which all earth's inhabitants, including us, depend. Part of a theological response to this state of affairs must be a willingness to change our images of God from those that sanction violence towards others and degradation of the earth's resources, to those that encourage us to relate nonviolently to each other, to the earth and to God.

It has also become clear that such image changing will require a 'willingness to change our minds and start again' by foregoing a religious perception of ourselves as belonging to a select number of humans chosen or delegated by God to have power over earth's life support systems and over all those dependent on them, whether human or more-than-human. The real impact of this way of regarding ourselves and, in effect, of having regard for ourselves alone, has been that we have abused those systems for our own benefit, with that benefit being computed, for the most part, in terms of monetary profit.

This has meant that that benefit has been, and is, calculated independently of the well-being of the whole Gaian community. That type of calculation presupposes our sole right to exploit earth's life systems for our immediate satisfaction, with that satisfaction largely measured now in terms of increased and ever-increasing consumption. If questioned, this presupposition is religiously defended on the grounds of our having a divine mandate to use the earth, or simply because we are able to do so because of our God-given intelligence.

These religious presuppositions have built on and been reinforced by images of God as Master, Lord and King of the universe, one exercising symbolic *punitive* power over the earth and its inhabitants in ways that have legitimated, indeed encouraged enmity rather than amity between peoples. And those images in turn have been used in the past, and continue to be used, as a religious licence for employing the utmost violence against those we designate as God's enemies. This conjunction of a capitalist system with a religious aura has persisted because of that system's ability to produce and reproduce surplus monetary value for certain social classes. Also, by selective use of biblical sources, it has produced its own symbolic universe, its own spirituality and its own religion. Within this symbolic universe, appropriation of land by violent means and its use for monetary profit is presented as a divine ordinance (Hinkelammert 1986: xiv).

Yet it is generally accepted that the word 'Gaia' is a collective noun, used by James Lovelock and throughout this book to signify a collective system or entity that may be imagined and analysed in various ways: but never in a way that separates us out from that collective whole. We cannot function outside it and ultimately, what works for its benefit works also for ours. But any specific benefit to us must be attained with consideration for that of the whole. Similarly, what we do against the good of the whole ultimately works against us also in this very complex, dynamic global system of interaction between living and non-living components.

It has also become clear that historic events have affected the evolution of the Christian symbolic universe and its conjunction with capitalist systems. They have also affected our activities and through their givenness, affected the whole of the collective living entity called Gaia. The evidence of our mutual assured vulnerability within Gaia is there for us to see in the event of climate change and its effects. It is no respecter of persons, species, ecosystems or lifestyles. Even though we, as individuals, cannot see all that is involved in this change, we are experiencing certain of its effects depending on our situation within the global weather system. From these, scientists can extrapolate a view of the whole, and our perception of that whole has been expanded by media coverage of and access to their findings. Public insistence by scientists from all disciplines on the fact that extreme weather events are associated with climate change is now, therefore, accompanied by political acknowledgments that we and our activities have played, and are playing, a pivotal role in that change. This in turn is having a transformative effect on our perception of being part of the collective entity and system signified by the name 'Gaia'.

The emergence of postcolonialist studies, together with movements for gender, racial and environmental justice, has also increased our sensitivity to the violent nature of original encounters between European and other cultures and to their enduring effects on other continents. Looking at this from a theological standpoint, interreligious dialogue, gender-based *and* postcolonial deconstruction and reconstruction of traditions and texts, ecological justice studies, postmodern philosophy and historical biblical criticism have contributed to a more broadly based religious consciousness, as has growing awareness of what remains of indigenous cultures. Christian images of God are also being transformed by scholarly excavation into the life, times and teaching of the historical Jesus.

One positive outcome of these latter changes has been the recovery of Jesus' uncompromising teaching about the love of enemies and with it the role of unconditional forgiveness, briefly discussed in the previous chapter. This *teaching* directly challenges doctrines of divinely sanctioned violence towards others of different racial, sexual, ethnic or religious identity, and by extension, violence towards others of different species identity. The remainder of this chapter will be devoted to taking a closer look at what Arendt called Jesus' discovery of the role played by forgiveness in the realm of human affairs.

The nature of forgiveness

After Arendt made this claim about Jesus, quoted in the epigraph to this chapter, she attended and reported on Eichmann's trial in Israel. There she had to face the reality of the demands made by forgiveness on an audience filled with European immigrants who, like herself, already 'knew by heart all there was to know' as witness followed witness and horror piled on horror before them (Arendt 1964: 8). Derrida also reflected on the complexities of this role as played out in such 'Theatres of Forgiveness' as the Nuremberg trials, the Truth and Reconciliation Commission in South Africa and the attempts in post-war France to purge itself of guilt after the Occupation by Germany in the Second World War. Of personal concern to him were French attempts to deal with the effects of the Algerian war (Derrida 2001: 27–58).

So there is nothing glib or superficial about the insights offered by both of them into what forgiveness requires of us: or about the role they propose it should play in the realm of human affairs. Nor, indeed, about their perception that it is an absolute requirement of those claiming to follow Jesus. Similarly, for Christians in today's world of ceaseless war, enemy-focused identity-formation and demands for personal and national security, it has become a *statu confessionis* to cultivate a world view based on a love of enemies with forgiveness as its active manifestation. For it is a point where Christians must make a clear decision and take a strong stand on how we relate to each other: hopeful of the ensuing consequences for ourselves and others and of a positive outcome from present conflicts. Such a commitment to forgiveness is a way of changing the quality of our relationships and frees us to begin something new that affects not only the realm of human affairs but ultimately the whole of earth's community of life on which we depend and of which we are part.

The insights Arendt offers into the nature and importance of forgiveness leave us in no doubt about the challenge this poses. For it involves a change of mind, heart and lifestyle without which we cannot free ourselves from the harmful effects of past events and are not, therefore, free to begin something new. The words stemming from *aphiemi* in the Greek New Testament are translated into English as 'for-give' just 47 times but are used in total (as in Matthew 6:12) 146 times. (Its etymological roots in 'giving' and 'gift' are clearer in the French translation: *par-donner* and *par-don*.) On other occasions it is translated by English words or phrases such as 'leave', 'let alone', 'release', 'give up', 'dismiss', 'forsake': in all of these the central idea is to 'let go', usually to let go of pain and hurt but also, and very importantly, of the desire to inflict pain or hurt in return. This not only expands and thickens up the meaning of the word 'forgive'. It also signals that forgiveness is a process: indeed a profound, painful and lengthy commitment and undertaking.

In this respect, forgiveness is the exact opposite of vengeance, which acts in the form of re-acting against an original trespass against us, so that far from putting an end to the consequences of the first misdeed, everyone remains bound to a violent process, permitting the chain reaction contained in every action to take an unhindered course. We are living in the midst of such a chain reaction and its predictable consequences under the slogan of 'the war against terrorism'. And in a different, but no less real, way with the climatic consequences of a reaction to the rise in carbon emissions that scientists call positive feedback. In contrast to these predictable outcomes, the act of forgiveness retains the chaotic capacity to create something new that can never be predicted. It is the re-action that acts in an unexpected way and so retains, though remaining a reaction, something of the original character of action.

> Forgiving, in other words, is the only reaction which does not merely re-act but acts anew and unexpectedly, unconditioned by the act that provoked it and therefore freeing from its consequences both the one who forgives and the one who is forgiven.
>
> (Arendt 1958: 241)

The role of forgiveness

I am using the term forgiveness here as Arendt does, with the strong New Testament Greek connotations of 'changing one's mind', 'releasing' and 'letting go'. All of these have enduring consequences for those involved, whether it be the one who forgives or the one forgiven. Therefore it requires that we be prepared to commit ourselves to constantly, mutually releasing each other from the destructive effects of what we do, thereby breaking the circle of vengeance and environmental destruction. But this is more than a simple exteriority from that circle. It is also now, for all of us in consumerist societies, a mutual release from the mindset of an economic fiscal cycle that keeps our relationships with others running along a Möbius strip of debt and credit, credit and debt.

This insight lay behind the biblically inspired Jubilee 2000 movement in the year 2000, when there was a concerted effort to have the overwhelming monetary indebtedness of 'Third World' nations remitted. It was intended to release them from that burden so that they could begin something new in the year of jubilee:

> [T]he land shall not be sold in perpetuity, for the land is mine ... And in all the country you possess, you shall grant a redemption of land ... And if your brother becomes poor and cannot maintain himself, you shall maintain him; as a stranger and a sojourner he shall live with you. Take no interest from him, or increase, but fear your God; that your brother may live beside you. You shall not lend him your money at interest, nor give him your food for profit ... And if your brother becomes poor and sells himself ... and if he is not redeemed ... then he shall be released in the year of jubilee, he and his children with him.
>
> (Leviticus 25: 23, 35–7, 54)

Even to start thinking along these lines requires a willingness to change our minds about what we owe and what is owed to us: and then to start thinking about what we all need to live well rather than what we want to possess. Actually doing this – and it is an essential part of the process of change – would free us from the desire to increase our possessions that continues to lay waste earth's resources.

> Without being forgiven, released from the consequences of what we have done, our capacity to act would, as it were, be confined to one single deed from which we could never recover; we would remain the victims of its consequences forever, not unlike the sorcerer's apprentice who lacked the magic formula to break the spell.
>
> (Arendt 1958: 237)

An understanding of forgiveness as 'mutual release' from either debt or credit that can serve as a prerequisite for action does not mean that the effects of what we have done or of the givenness of events are wiped out. It does mean that, through the act of forgiveness, we are free to work imaginatively and long term at transforming those effects. The jubilee tradition envisioned economic forgiveness and change over a fifty-year cycle. The Jubilee 2000 movement, with its logo of broken chains, was inspired by this vision of a change in human affairs releasing people and land enmeshed in debt and by freeing them from it, positively affecting both. That is the relational model behind Jesus' command about not charging interest.

> Jesus says: If you have money, do not lend out at interest. Rather, give it to the one from whom you will not get it back.
>
> (Gospel of Thomas 95 in Robinson *et al.* 2002: 85)

Caputo is thinking of such absolute commands when he says that forgiveness is the ultimate release from all (worldly) economies, from every economic tie.

Forgiveness loosens the circle of credit and debt, not only from the debt that chains the other with the tie of my calculated gift, but also from the debt that makes my relation to the other one of debt.

(Caputo 1997: 227)

Here I want to stress, with Stăniloae, that this conjunction of freedom with action is the prerequisite for being able to transform the world for the better through our labours and, if you will, return it to God as a free gift. This attitude to labour may sound utopian: yet the Jubilee texts are realistic about the fact that our brothers and sisters, then and now, yesterday and today, are forced to sell themselves into bonded labour in order to survive. The survival of exploited workers now depends on decisions made by capitalists in order to satisfy their desire to accumulate surpluses for unrestrained consumption.

The harsh reality of this makes it all the more scandalous that for some within Western Christian traditions, our labour is perceived in terms of payment for a debt incurred by sin: one imposed on us by God. That surely must change, and thereby change our relation to the God of Jesus. For to understand and to accept his teaching about forgiveness requires a refusal to envision his life and, in particular, his death, as an amortization of an alleged debt incurred by 'Adam's sin'. Nor does his teaching support an image of God drawn in such punitive economic terms.

If it did, it would go against what Arendt discovers in and attributes to Jesus: that forgiveness is the ideal model for restoring relationships within and between communities and between us and God. For it creates amity rather than enmity, releasing us from an obligation to re-act to God as one who expects a return for the gift of the world. Instead, it invites us to act freely and thereby respond positively to its character as gift.

For it is characteristic of giving that it requires that we forgive and are for-given *freely*. It also requires us to love and be loved *freely*, whether the object of our love is God, or the Samaritans and Canaanites in our lives. And to be loved freely by them. So finding and fostering the ability in ourselves to forgive makes it possible for us to initiate new and nonviolent relationships in the realm of human affairs that contribute to the well-being of the whole Gaian community of life.

Our bodies signify everything that is vulnerable within that community, including our mutual (assured) vulnerability to the effects of climate change. But they also signify all that is lovable and are, for the same reason, mendable, healable, transformable, capable of undergoing change, of changing our mind and heart and transforming our disposition (Caputo 2006: 131–3). Forgiveness releases our capacity for change and enables us to act for the good of the whole by letting us see how to begin acting differently.

The role of forgiveness in human communities

Arendt insists that although Jesus discovered the role of forgiveness in the realm of human affairs within what may be perceived as solely a religious context, and articulated it in what we read and hear as religious language, this is not a reason to take it any less seriously. She sees his discovery, rightly, not as primarily related to the Christian religious message but as springing from experiences within the small and closely knit community of Jesus' followers bent on challenging, by their very existence, the public authorities in Israel. Remembering Robinson's and Auerbach's comments on the cultural, economic, social and religiously deprived character of that community vis-à-vis such authorities, it represents one that, in Galtung's terms, was peripheral to the Centre in the Periphery nation, Israel. Arendt does not see any awareness in the Jewish religious authorities at that centre that forgiveness might be a necessary corrective to the inevitable damage resulting from their actions. In other words, it was not a predictable but a new, chaotic, creative response by Jesus.

She does, however, discern such an awareness at the Centre of the Roman Empire in its principle to spare the vanquished – a wisdom, she says, entirely unknown to the Greeks. It is evidenced in the occasional Roman practice of commuting the death sentence, an instance of which is recorded in John's Gospel where Pilate employs it in relation to Barabbas (John 18: 38–9). It is, therefore, decisive for her that Jesus maintains, *against* the authority of 'the scribes and Pharisees', that it is not only God who has the power to forgive. She notes that this is stated emphatically in Luke 5: 21–4 (see also Matthew 9: 4–6, Mark 12: 7–10), where Jesus performs a miracle to prove that 'the Son of Man has power upon earth to forgive sins': the emphasis being on 'upon earth'. It is his insistence on having 'the power to forgive' even more than his performance of miracles that, she says, shocks the people. So that those who sat at the table with him began to say within themselves, 'Who is this that forgives sins also?'

The power to forgive

It is also decisive for Arendt that this power is not seen to derive from God and then mediated through human beings, as though God, not us, is still the one who forgives. On the contrary, that power must be mobilized by us here and now towards each other before we can hope to be forgiven by God as well. God must wait, so to speak, on our willingness to forgive: a willingness that must be mobilized as frequently as needed. To highlight this essential and continuous earthly interaction she analyses the text where Jesus lays down the protocols for forgiving:

> If your brother sins, rebuke him, and if he repents, forgive him, and if he sins against you seven times in the day, and turns to you seven times and says: 'I repent', you must forgive him.
>
> (Luke 17: 3–4)

This is the Revised Standard Version of the passage. Expanding on what was said earlier, Arendt takes account of certain connotations of New Testament Greek that underlie three key Greek words in the text – *aphiemi, metanoein* and *hamartanein* (Zerwick 1966: 249). She translates them accordingly and notes that the original meaning of *aphiemi* is 'dismiss', 'release', 'let go', all implicit in 'forgive' though now almost forgotten. *Metanoein* means 'change of mind' and – since it serves also to render the Hebrew *shuv* – 'return', 'trace back one's steps', she suggests these active responses should be stressed rather than 'repentance' with its psychological emotional overtones. Finally, the word translated as sin, *hamartanein,* is very well rendered by 'trespassing' in so far as it means rather 'to miss the mark', 'to fail and go astray', meanings that can be missed by simply using the words 'to sin'. So what Jesus requires of us could also be rendered as follows:

> And if he trespass against thee ... and ... turn again to thee, saying, *I changed my mind,* thou shalt *release* him.
>
> (Arendt 1958: 240)

Arendt points out that Jesus' formulations are still more radical. He does not say that we are supposed to forgive because God forgives and so we must do 'likewise', *but* 'if you from your heart forgive, *God* shall "do likewise"' (Matthew 6: 12, 18: 35; and see Mark 11: 25). God forgives us our debts, '*as we* forgive our debtors' (Arendt 1958: 239).

Here she anticipates Caputo's highlighting of the 'hinge' of Jesus' logic, the *hos kai:* 'Forgive us *as indeed* we forgive others' (Caputo 1997: 226). But this radical demand of God says something even more radical about God's power to forgive and about power within the kingdom of the powerless. For Jesus makes the use of God's power to forgive dependent on our use of it. Is this not the image of a 'reticent' god who gives everything away, including the awesome power to forgive sins? And with it, the power to punish them? Is this not also the image of a god who does not exact vengeance on our enemies but instead, commands us to love them?

The power to forgive is ours. Its exercise has personal, collective, structural and symbolic effects within the kingdoms of this world that can, if we wish to act in accordance with Jesus' 'hard sayings' about it, transform the poisonous effects of personal, structural, ecological and symbolic violence. This is because forgiveness presupposes relationships to others within a community where we renounce getting even, or squaring accounts, or rendering reasons equal on each side, or calculating the amount needed for retribution, or paying back in kind. It arises directly out of the necessity of living with others and from the will to live with them as peaceably, lovingly and generously as possible.

> The spirit of giving is not *in* us, but *between* us. It is not fully expressed or made manifest in what is given but in the process of giving. Ideally, it enables us to respond to each other, to relate to each other as 'You'; without simply desiring to possess or use each other – or what is given.
>
> (Buber 1970: 153–6)

Buber links the mode of the relationships between us with the mode of our relationships with what is given. Both are to be marked by releasing ourselves from the desire to possess or use. But it is in our attending to the nature and role of the responses *between* us, and between us and the gifts we receive, that we realize this ideal. No individual body can forgive herself. Forgiving and the relationship it establishes is always an eminently interpersonal, interactional process in which *what* is done is forgiven for the sake of the relationship between both the one who has to be forgiven and the one who has to forgive. For this flows from and restores the possibility of loving the other freely. This too was clearly recognized by Jesus.

> Her sins, which are many, are forgiven: for she loved much: but he who is forgiven little, loves little.
>
> (Luke 7: 47)

For love possesses an unequalled power of self-revelation and an unequalled clarity of vision for the disclosure of *who* the loved person may be, with her qualities and shortcoming no less than her achievements. Here, as in action and speech generally, we are dependent upon others, for we appear to them with a distinctness that we do not have in regard to ourselves. If we remained closed in upon ourselves we would never be able to forgive ourselves for failings or transgressions, because we would lack the experience of the person for the sake of whom one can forgive (Arendt 1958: 242–3).

That is why forgiveness is an event within the kingdom of God *on earth* and not an individual experience. It is an event that begins to start something new in a place and between people here and now: in the only space where we can love and hurt each other. It belongs to 'the very quintessence of the human condition' (Arendt 1958: 2). This view of forgiveness itself belongs within Arendt's positive view of earth, for she made *amor mundi,* love of the world, the foundation for ethical and political action. It was not, for her, a romantic or sentimental attitude to a world where all is daffodils and songbirds. This could hardly be possible considering her own history as a twentieth-century, German-born Jew.

In contrast to the denigration of earthly birth in favour of a life in heaven, she emphasized respect for and fostering of 'natality', of the new birth of human beings and of the hope they bring into the world that is our home. For Arendt, to be dedicated to the world meant among other things not to act as though we were immortal. Rather, our commitment to the world entails dedication to the 'constant influx of newcomers born into the world as strangers' since the future of the world is dependent on them after we are gone (Jantzen 1998: 152).

Saying this she went against the grain of *contemptus mundi,* the contempt for the world that, says Grace Jantzen, has characterized the Western religious symbolic system and the secularism that grew out of it. An emphasis on birth, a dedication to the well-being of newcomers into the world and on natality in all beings is not, however, a denial of death or a pretence that death does not exist. Nor does it deny the fact that a sense of commitment to newcomers into the world means that the labour involved in dedication to nurturing their lives is necessarily painless, congenial or immediately rewarding.

Love of the earth does, however, pro-claim a love of life: and beyond our individual lives a love of the life of the world and of the lives of those to come for whom it will be a home. Love of the world is not to be characterized as emotion but as a deliberate choice. In Arendt's terms: '*Amo: Volo ut sis* – I love you: I will that you be.' Similarly she deplored action that treats the earth as though it were nothing but an inexhaustible supply of material for our gratification. It is, remarked Jantzen presciently, this rejection of limits that is bringing about the increasingly rapid destruction of the habitable world (Jantzen 1998: 152–4).

Can Gaia forgive us?

In the light of what has been said about the role of forgiveness in the realm of human affairs and in the kingdom of God on earth, the question now is: 'Can Gaia forgive us?'

At best, the question itself implies a genuine awareness that we have inflicted harm on Gaia, that we have trespassed on and harmed the sacred ground common to all life on earth. It also acknowledges that we have failed, gone astray from the right way of living in the world and from the spirit of giving that enables us to respond to the earth's gifts without simply desiring to use or possess them. For if so, we have 'missed the mark' by aiming to accumulate possessions for ourselves rather than focusing on the long-term good of the whole.

Following on from the definitions and discussion of this chapter, if we want Gaia to release us from the effects of what we have done, then we have to retrace our steps and find out where we went astray. One wrong road we have taken has been that which has led to the global dominance of the present capitalist economy and its emphasis on transforming nature into short-term monetary profit that drives the engines of consumption and waste of resources. Within the present context of highly industrialized and technologized societies in the G8 countries, our labour and Gaia's resources are consistently downgraded, ignored or exploited by corporations who neither recognize nor act upon moral criteria that would prevent them from harming others.

Indeed, they are legally compelled by pragmatic concern for their own interests to cause harm when the benefits to shareholders of doing so outweigh the costs. The technical jargon of economics describes that harm as an 'externality', that is, it affects a third party who has not consented to or played any role in carrying out a transaction. When an 'externality' badly affects people and the environment, executives have by law neither obligation nor authority to consider those effects unless they might negatively affect the corporation itself (Bakan 2004: 60f.).

This 'externalizing machine', as Bakan calls it, is an apt metaphor for a mind-set that sees us and our activities as in some way external to Gaia. The reality is, of course, that nothing in our lives or actions is external to the collective living entity of Gaian life support systems that ineluctably connect our lives and activities with those of all other beings on earth.

Therefore, the injuries we inflict on those systems have a dual aspect: the collective or structural one attributable to us as a species, and that resulting from our actions as individuals. Within the community of life on earth both of these contribute, or not, to the well-being of the whole. Therefore, we cannot externalize the harm we do to Gaia from the harm we do to each other. Here again the 'hinge' of Jesus' saying links together our individual need for and expression of forgiveness and the need for and expression of forgiveness within Gaia. Gaia forgives us, or not, *as indeed* we forgive each other.

This brings us back to another aspect of our relationship with God, and by analogy with Gaia, that emerges from Jesus' teaching on forgiveness. God's primary concern goes way beyond physically harming another being. Giving way to anger or insulting another person will make our relationship with God untenable. This logic emerges when, in the context of going to the temple as part of religious observance, Jesus lays down the conditions for making our gift there acceptable. Attention is focused not on our relationship with God but on our relationships with each other.

> If you are offering your gift at the altar and there remember that your brother or sister has something against you, leave your gift there before the altar and go: first be reconciled with your brother or sister and then come and offer your gift.
>
> (Matthew 5: 21–4)

Not only are we to forgive. We are actively to seek forgiveness for any hurt we have caused to each other. As Jesus says, God is concerned first and foremost with our relationships to each other.

Gaia's concerns, too, are not with a system set out as a flow chart, nor with separate entities that can be listed, nor with a theory and certainly not with the mythological goddess of antiquity. Gaia's concerns are with the sum and product of the uncountable and infinitely complex intra-Gaian relationships that contribute to and constitute the well-being of the whole. The character of those relationships determines not only the givenness of the climate in future but the givenness of all future life on earth.

Bibliography

Arendt, H. (1958) *The Human Condition*. Chicago, University of Chicago Press.

Arendt, H. (1964) *Eichmann in Jerusalem: A Report on the Banality of Evil*. London, Penguin.

Arneil, B. (1996) *John Locke and America: the Defence of English Colonialism*. Oxford, OUP.

Auerbach, E. (1953) *Mimesis: the Representation of Reality in Western Literature*. Princeton, Princeton University Press.

Badiou, A. (2001) *Ethics: An Essay on the Understanding of Evil*. London; New York, Verso.

Bagchi, D. V. N. (2000) 'Reformation'. *The Dictionary of Historical Theology*. T. Hart (ed.) Carlisle, Paternoster Press: 462–6.

Bakan, J. (2004) *The Corporation: The Pathological Pursuit of Profit and Power*. London, Constable.

Baldwin, J. (1990) *Notes of a Native Son*. Boston, Beacon Press.

Barnes, P. (2006) *Capitalism 3.0: A Guide to Reclaiming the Commons*. San Francisco, Berrett-Koehler.

Barrow, J. D. and F. J. Tipler (1986) *The Anthropic Cosmological Principle*. Oxford, OUP.

Bauman, W. A. (2007) 'Creation ex Nihilo, Terra Nullius and the Erasure of Presence.' *Ecospirit: Religions and Philosophies for the Earth*. L. Kearns and C. Keller (eds). New York, Fordham University Press: 353–72.

Berry, T. (2006) 'Loneliness and Presence.' *A Communion of Subjects*. K. Patton and P. Waldau (eds). New York, Columbia University Press: 5–10.

Betts, R. (2007). 'Human-Caused Climate Change.' *Earthy Realism: The Meaning of Gaia*. M. Midgley (ed.). Exeter, Imprint Academic: 39–45.

Blond, P. (ed.) (1998) *Post-secular Philosophy: Between Philosophy and Theology*. London, Routledge.

Blumenberg, H. (1987) *The Genesis of the Copernican World*. Cambridge, Massachusetts, The MIT Press.

Borg, M. J. and J. D. Crossan (2007) *The Last Week: What the Gospels Really Teach about Jesus's Final Days in Jerusalem*. San Francisco, HarperSanFrancisco.

Buber, M. (1970) *I And Thou*. Edinburgh, T & T Clark.

Caputo, J. D. (1997) *The Prayers and Tears of Jacques Derrida*. Bloomington, Indiana, Indiana University Press.

Caputo, J. D. (2006) *The Weakness of God: A Theology of the Event*. Bloomington, Indiana, Indiana University Press.

Caputo, J. D. (2007) 'After the Death of God.' *After the Death of God*. J. W. Robbins (ed.). New York, Columbia University Press.

Caputo, J. D. and M. J. Scanlon (eds) (1999) 'Introduction: Apology for the Impossible: Religion and Postmodernism.' *God, the Gift, and Postmodernism*. Caputo, J. D. and M. J. Scanlon (eds). Bloomington, Indiana, Indiana University Press.

Carson, R. (1962) *Silent Spring*. London, Penguin.

Compier, D. H. (2007) 'Jean Calvin.' *Empire: The Christian Tradition*. Kwok P-L, D. H. Crompier and J. Rieger (eds). Minneapolis, Fortress Press: 215–27.

Crossan, J. D. (1994) *Jesus, A Revolutionary Biography*. New York, HarperSanFrancisco.

Crossan, J. D. (1999) 'Our Own Faces in Deep Wells: A Future for Historical Jesus Research.' *God, the Gift and Postmodernism*. J. D. Caputo and M. J. Scanlan. Bloomington, Indiana, Indiana University Press 282–310.

Crossan, J. D. and J. L. Reed (2001) *Excavating Jesus: Beneath the Stones, Behind the Texts*. London, SPCK.

Deane, S. (1983) *Civilians and Barbarians*. Belfast, Field Day Theatre Company.

Deleuze, G. (1990) *The Logic of Sense*. London, Continuum.

Deleuze, G. and F. Guattari (1988) *A Thousand Plateaus*. London, The Athlone Press.

Derrida, J. (1995) *The Gift of Death*. Chicago, University of Chicago Press.

Derrida, J. and J-L. Marion (1999) 'On the Gift: A Discussion between Jacques Derrida and Jean-Luc Marion.' *God, the Gift and Postmodernism*. J. D. Caputo and M. J. Scanlan. Bloomington, Indiana, Indiana University Press: 54–79.

Derrida, J. (2001) *On Cosmopolitanism and Forgiveness*. London, Routledge.

Douglas, M. (2000) *Leviticus as Literature*. Oxford, OUP.

Drees, W. B. (2002) *Creation: From Nothing until Now*. London, Routledge.

Duchrow, U. and F. J. Hinkelammert (2004) *Property for People: Not for Profit: alternatives to the global tyranny of capital*. London; New York, Zed Books.

Dussel, E. (1990) 'The Real Motives for the Conquest' *Concilium* 6: 30–46.

EOLSS, The Encylopaedia of Life Support (http://www.eolss.net)

Franklin, U. (1990) *The Real World of Technology*. Toronto, CBC Enterprises.

Funk, R. W. (1990) 'In the Heart of America: Redeemer Figures and Mythic Spaces.' *The Fourth R*. 3: 1–4.

Funk, R. W. (1993) 'The Gospel of Jesus and the Jesus of the Gospels.' *The Fourth R*. 6: 3–10.

Funk, R. W. (2000) 'The Once and Future Jesus.' *The Once and Future Jesus*. Santa Rosa, Polebridge Press: 5–27.

Funk, R. W. (2002) *A Credible Jesus: Fragments of a Vision*. Santa Rosa, Polebridge Press.

Funk, R. W. and W. Roy (eds) (1993) *The Five Gospels*. New York, Macmillan.

Furlong, A. (2003) *Tried for Heresy: A 21st-Century Journey of Faith*. Alresford, O Books.

Galston, D. (2007) *Liturgy in the Key of Q*. Westar Institute Spring Conference, Santa Rosa, Westar.

Galtung, J. (1971) 'A structural theory of imperialism.' *Journal of Peace Research*. 8: 81–117.

Gleick, J. (1998) *Chaos: the Amazing Science of the Unpredictable*. London, Vintage.

Gutierrez, G. (1996) 'Option for the Poor.' *Systmeatic Theology.* J. Sobrino and I. Ellacuria (eds). London, SCM Press.

Haraway, D. (1997) *Modest_Witness@Second_Millennium.* London, Routledge.

Henson, R. (2006) *The Rough Guide to Climate Change.* London, Rough Guides Ltd.

Hinkelammert, F. J. (1986) *The Ideological Weapons of Death: A Theological Critique of Capitalism.* Maryknoll New York, Orbis.

James, W. (1960) *The Varieties of Religious Experience.* London and Glasgow, Collins.

Jantzen, G. (1998) *Becoming Divine: Towards a Feminist Philosophy of Religion.* Manchester, Manchester University Press.

Jennings, T. W. (2007) 'John Wesley.' *Empire: The Christian Tradition.* P-L. Kwok, D. H. Compier and J. Rieger (eds). Minneapolis, Fortress Press: 257–70.

Keller, C. (2003) *Face of the Deep: A Theology of Becoming.* London, Routledge.

Keller, C. (2005) *God and Power: Counter-Apocalyptic Journeys.* Minneapolis, Fortress Press.

Kelly, J. N. D. (1950) *Early Christian Creeds.* London, Longmans, Green and Co.

Kennelly, B. (2004) *Familiar Strangers.* Northumberland, Bloodaxe Books.

Klinkenborg, V. (2007) 'Millions of Missing Birds, Vanishing in Plain Sight', *New York Times,* 19 June.

Koyré, A. (1957) *From the Closed World to the Infinite Universe.* Baltimore, The Johns Hopkins University Press.

Kwok, P-L. (2005) *Postcolonial Imagination and Feminist Theology.* Louisville, Westminster John Knox Press.

Leakey, R. and R. Lewin. (1996) *The Sixth Extinction: Biodiversity and Its Survival* London, Weidenfeld & Nicolson.

Liu, J., T. Dietz, *et al.* (2007) 'Complexity of Coupled Human and Natural Systems.' *Science* 317(5844): 1513–16.

Long, A. (1992) *In a Chariot Drawn by Lions.* London, The Women's Press.

Lorenz, E. N. (1995) *The Essence of Chaos.* Seattle, University of Washington Press.

Lovelock, J. (1991) *Gaia, The Practical Science of Planetary Medicine.* London, Gaia Books.

Lovelock, J. (1995) *The Ages of Gaia.* Oxford, OUP.

Lovelock, J. (2007) Foreword. *Earthy Realism.* M. Midgley (ed.). Exeter, Imprint Academic: 1–2.

Lovelock, J. (2007) *The Revenge of Gaia.* London, Penguin.

Margulis, L. (1986) *Microcosmos: Four Billion Years of Microbial Evolution.* Berkeley, University of California Press.

Margulis, L. (1998) *The Symbiotic Planet.* London, Weidenfeld & Nicolson.

Marion, J-L. (1991) *God without Being.* Chicago, University of Chicago Press.

Marion, J-L. (1998) *Reduction and Givenness: Investigations of Husserl, Heidegger and Phenomenology.* Evanston, Northwestern University Press.

Marion, J-L. (1999) 'On the Gift: A Discussion between Jacques Derrida and Jean-Luc Marion.' *God, The Gift and Post-Modernism.* J. D. Caputo and M. J. Scanlan. Bloomington, Indiana, Indiana University Press: 54–79.

Matheson, P. (2006) 'The Reformation.' *The Blackwell Companion to the Bible and Culture.* J. F. A. Sawyer. Oxford, Blackwell: 69–84.

McNeill, J. R. (2000) *Something New Under The Sun: An Environmental History of the Twentieth-Century World.* New York; London, W. W. Norton.

Merchant, C. (1980) *The Death of Nature: Women, Ecology and the Scientific Revolution.* New York, Harper and Row.

Miller, C. (2000) *The Gift of the World*. Edinburgh, T &T Clark.

Montefiore, C. G. and H. Loewe (eds) (1974). *A Rabbinic Anthology*. New York, Schocken Books.

Myerson, G. (2001). *Ecology and the End of Postmodernity*. Cambridge, Icon Books.

Nelson-Pallmeyer, J. (2001) *Jesus Against Christianity: Reclaiming the Missing Jesus*. Harrisburg, Pennsylvania, Trinity International.

Nelson-Pallmeyer, J. (2007) 'Another Inconvenient Truth: Violence within the "Sacred" Texts.' *The Fourth R*. **20**: 9–15.

Niditch, S. (1985) *Chaos to Cosmos: Studies in Biblical Patterns of Creation*. Atlanta, Georgia, The American Scholars Press.

Nineham, D. (1993) *Christianity Mediaeval and Modern*. London, SCM Press.

O Crualaoich, G. (2003) *The Book of the Cailleach: Stories of the Wise-Woman Healer*. Cork, Cork University Press.

Orion magazine, found at: http://hubble.nasa.gov and http://www.hubblesite.org

Pagels, E. (1990) *Adam, Eve and the Serpent*. London, Penguin.

Patterson, S. (1998) *The God of Jesus*. Harrisburg, Pennsylvania, Trinity Press International.

Posey, D. A. (ed.) (1999) *Cultural and Spiritual Values of Biodiversity: United Nations Environment Programme*. London, Intermediate Technology Publications.

Prigogine, I. and I. Stengers (1985) *Order out of Chaos*. London, Flamingo.

Primavesi, A. (1998). 'Gaia Theory and Environmental Policy.' *Spirit of the Environment*. D. E. Cooper and J. A. Palmer. London, Routledge: 204.

Primavesi, A. (2000) *Sacred Gaia*. London; New York, Routledge.

Primavesi, A. (2003) *Gaia's Gift: Earth, Ourselves and God after Copernicus*. London; New York, Routledge.

Primavesi, A. (2004) *Making God Laugh: Human Arrogance and Ecological Humility*. Santa Rosa, Polebridge Press.

Primavesi, A. (2007) 'The Preoriginal Gift – and our Response to It.' *Ecospirit: Religion, Philosophy and the Earth*. L. Kearns and C. Keller, (eds). New York, Fordham University Press.

Richards, J. F. (2003) *The Unending Frontier: An Environmental History of the Early Modern World*. Los Angeles, University of California Press.

Rilke, R. M. (1975) *Poems from the Book of Hours*. New York, New Directions Publishing Coporation.

Robbins, J. W. (ed.) (2007) *After the Death of God*. New York, Columbia University Press.

Robinson, J. M. R. (2005) *The Sayings Gospel Q: Collected Essays*. Leuven, Leuven University Press.

Robinson, J. M. R. (2007) *Jesus: According to the Earliest Witness*. Minneapolis, Fortress Press.

Robinson, J. M. R., P. Hoffmann and J. S. Kloppenborg (eds) (2002) *The Sayings Gospel Q in Greek and English*. The International Q Project. Minneapolis, Fortress Press.

Rosenzweig, F. (1985) *The Star of Redemption*. New York, University of Notre Dame Press.

Rowbotham, M. (1998) *The Grip of Death: A study of modern money, debt slavery and destructive economics*. Charlbury, Jon Carpenter.

Rudwick, M. J. S. (2005) *Bursting the Limits of Time: The Reconstruction of Geohistory in the Age of Revolution*. Chicago, University of Chicago.

Schrödinger, E. (2000) *What is Life?* Cambridge, CUP.

Schwartz, R. (1997) *The Curse of Cain: The Violent Legacy of Monotheism*. Chicago, Chicago University Press.

Schweitzer, D. (2007) 'Jonathan Edwards.' *Empire: The Christian Tradition*. P-L. Kwok, D H. Compier and J. Rieger (eds) Minneapolis, Fortress Press: 243–56.

Segundo, J. L. (1985) *Theology and the church*. Minneapolis, Winston Press.

Snyder, G. (1990) *The Practice of the Wild*. New York, Farrar, Straus and Giroux.

Sobrino, J. (1985) *The True Church and the Poor*. London, SCM.

Suzuki, D. M. (1997) *The Sacred Balance: Rediscovering our Place in Nature*. Vancouver, Greystone Books.

Thompson, J. N. (2006) 'Mutualistic Webs of Species.' *Science*. **312** (5772): 372–3.

Tudge, C. (2005) *The Secret Life of Trees*. London, Penguin.

Volk, T. (2002) *What is Death? A Scientist Looks at the Cycle of Life*. New York, John Wiley and Sons.

Ware, T. (1984) *The Orthodox Church*. London, Penguin.

Weber, M. (1930) *The Protestant Ethic and the Spirit of Capitalism*. London, Routledge.

Weber, M. (1965) *The Sociology of Religion*. London, Methuen.

Weeden, T. J. (2007) *Excavating the Galilean Jesus Movement: Its Lost Gospel, Its Lost Tradition, Its Lost Jesus*. Westar Institute Spring Seminar. Miami, Westar.

Weil, S. (1951) *Waiting on God*. London, Routledge.

Weil, S. (2003) *Simone Weil on Colonialism: an ethic of the other*. Lanham, Maryland, Rowman and Littlefield.

Weil, S. (2005) *The Iliad, or the Poem of Force*. New York, New York Review of Books.

Whitehead, A. N. (1925) *Science and the Modern World*. New York, Free Press.

Wittgenstein, L. (1961) *Tractatus Logico-Philosophicus*. London, Routledge & Kegan Paul.

Worster, D. (1977) *Nature's Economy: A History of Ecological Ideas*. Cambridge, CUP.

Zerwick, M. (ed.) (1966) *Analysis Philologia Novi Testamenti graeci*. Rome, Pontifical Biblical Institute.

Index

Aboriginal Australians 46
accountability 12, 16, 31–2
acquisitiveness 42
actions, consequences of 14–15
actualization 19–20, 80
Adam, sin and death 71–2, 85–86, 135
adama (earth creature) 75
'Agenda 21' 78
Algerian war 132
aliens, humans as 11, 24, 30
ambiguity of gift events 69
America, discovery of 28–30, 38, 74–75;
    Locke and colonial policies 45
Amerindians 45, 46
amor mundi (love of the world) 47,
    138–9
Amsterdam Declaration 9–10, 14
ancestry and inheritance 68
ancestry of life, consequences of 34–5
Anselm 16
Anthropic Cosmological Principle 60,
    62–4, 72
aphiemi (let go, forgive) 132–3, 136–7
Apollo-Gaia Project 6
Apostles' Creed 109–10
Aquinas, T. on the purpose of labour 41
arbeit macht frei 43
Archimedean point 11, 22–3, 55–6,
    58–9
Archimedes 11, 22–3, 55–9, 62, 64
Arendt, H. 2–3, 12–13, 19, 28, 29–30,
    38–9, 42–4, 48, 53–65, 130, 132–9;
    on alienation 43; on Jesus' birth 55,
    57
Aristotle: on origin of money 44–5; on
    two economies, 44
arrogance 71–2
asceticism in-the-world 39; other-
    worldly, 39

astrophysical worldview 57
atomic bombs 11
Audubon Society 33
Auerbach, E. 128–9, 135–6
Augustine 85–8, 96–7, 103–4; The City
    of God 95–6; the fall of Rome 96;
    his view of Earth 85, 87–8, 94; on
    sin and death 85; on the gift of Jesus
    103–4
Auschwitz and the gift of bread 86–7,
    98
axis of evil 31

Badiou A. 20–21
Bagchi D. V. N. 38
Bakan J. 139
Baldwin J. 18
Bali 5, 78–9
banks and the ownership of money
    51–2
baptism 38–9, 119
barbarians and civilians 73–4
Barmen Declaration and the Confessing
    Church 53
Barrow J. 62–3, 72
Bauman, W. on Locke and the 'state of
    nature', 47
bean chaointe 72
bean feasa 72–3
bean ghluine 72–3
Beveridge, A. J. 18
Bhagavad Gita 61–2
Bible 2, 13, 16–17, 21, 24, 26–7, 37–9,
    42, 45, 53, 67, 73, 75–6, 82, 91–2,
    108–9, 119, 124–5, 131–2; effects
    of translation on 38–9; Luther's
    translation of 38–9
biodiversity 9, 14, 30, 42, 84
biofuels 78–9

biology 22–3
birds, decline of 33, 35, 45, 87, 97
Blake W. 82
blessing of food 87
Blumenberg, H. 56
bodies-in-the-making, humans as 34
bohu (emptiness in Genesis) 1–2, 6–7, 18, 20–21
Bollkaemper, J. 104
Bonhoeffer, D. 14, 53; status confessionis 53
Borg, M. and Crossan, J. D. on Jesus' passions 106–7
bread and stones in Q 127
British Empire 48
Bruno, G. 57, 60
Buber, M. 17, 36, 81, 92–3, 102, 137–8

Cailleach Bheara 77–8
Calvin, J.: on the acquisition of wealth 40; on the poor 43
Calvinism 39
Camara, H. 127–8
capitalist property 49
capitalist system 37, 131
Caputo, J. D. 1–4, 17, 20, 21, 25, 83, 87, 98–9, 101, 104–6, 119, 122–4, 134–7; on God's kingdom 104–6, 123; on kingdoms 101; on pure gift 123
Carson, R. 11
chaos 6–7, 18, 47–50, 58–9, 133–36
chemistry 21–2, 29
children 14–15, 36–7, 48–49, 52, 78, 106, 108, 118–21, 132
chosen people, on Israel as 13, 32, 41, 75, 114, 124, 130
Christendom 12–14, 26, 30–31, 89–90, 98, 115–16; as grand narrative 12–14, 26, 30–31, 89–90, 98, 115–16
Christian imperialism 75; colonialism and sexual violence 75
Christian violence 15, 115
Christianity: metamorphosis of 121–22
Church 12–13, 38, 40–42, 52–3, 97, 109–20, 124; its liturgy's primary focus 109; policy 120
civilians and barbarians 73–4
civilisation 13, 74
Clean Development Mechanism: effects on women 78–9

climate change 1–17, 24–5, 27–8, 30–37, 39–40, 49–53, 56–9, 64, 69–70, 73, 78–9, 83–6, 89–92, 94, 97, 111, 117, 121–2, 126, 129–31, 135; changes in world view brought about by 11, 14–15, 35, 58–9, 90–91; market-based solutions 78–9; a new context for human history 130; and our image of God 53, 89, 92, 130; and theological paradigm change 4, 90–91; and theology 3, 6, 12–15, 31, 79, 122, 126, 129; who bears the brunt of 129; women's reaction to 78
climate: the planetary unifying factor 126
climates: theological and cultural 91
closed world to infinite universe 57, 59–60, 62
Clotho, Lachesis, Atropos and the thread of life 84–5
Cold War 61, 76, 127–8
colonial exploitation 36, 42
colonialism 30, 75–7; Christian 30, 75; Christian imperialism and sexual violence 75
colonization 28–30, 42, 47–8, 73–8, 92; legitimations of 47–8; the legacy of 42, 73–7; violence done to women 76
Columbus, C. 30
commodities: as objects 50; as subjects 50; as players in the economy 50
commodity and producer 50
common environmental goods 52–3
commons 50–51
Confessing Church 53
consciousness 10, 12, 24–5, 33, 42–3, 64–5, 96, 110–11, 131–2; and delusion, 24–5
Constantine 12, 112–16, 119–18; and the imperial takeover of Christianity 12, 112–16, 119–20
Constantinople and Rome 89–10, 95–6, 105, 109, 114–15
consumption 6, 32–3, 39–40, 50–52, 79, 81, 91, 130–31, 135, 139
contamination of soil 12
context for giving 82–3
Copernican principle 63; world view 59
Copernicus 56–9
cosmic present 18–19
cosmos 17, 59

Council of Nicea 112-14
creation 1–2, 12, 16–27, 38–9, 46, 59,
    67, 70–1, 75, 80–1, 120–1; as
    seminal event 1–2, 15–27, 70–1,
    80–1; its situatedness 19
credal imperialism: lasting effects of 114
creeds and imperialism 110–112; Jesus'
    history missing in 108; the God of
    109
Crossan, J. D. 104–6, 113–16, 123–4;
    on historical Jesus resources 116; on
    the Council of Nicea 113–14
crucifix 110
crucifixion and images of God, 109
crusades: an era of Christian imperialism
    114–15, 115; and Jesus' command
    to love enemies 115–16

Deane, S. on the colonization of Ireland
    72–6
death: and Gaia 66–7, 70–71, 84–7; as
    punishment for one man's sin
    84–85; Christian views of 13, 70–1,
    84, 106, 109–10, 113, 115–16, 134
Deleuze, G. 2, 18–19, 67–70
Derrida, J. 3–4, 82–4, 99–100, 102,
    104–5, 122, 131; on gift 100,
    104–5, 122; on laws of God's
    kingdom 99, 102; on the living
    thread of giving 82–84, 100
desertification 8-9, 11, 35, 41
Digges, T. 59
dinosaurs 2–3
Diogenes 88–9
discourse: scientific 9, 23
distribution of goods unequal by divine
    providence 42
divine economy 36, 79, 86, 106
divine impartiality 117–20
Doerr, A. 57-8
domination: postcolonial 30
Douglas, M. 25–6
Drees, W. 17
Duchrow, U. 43–46, 48–52
Dussel, E. 29–30, 30
dynamics of giving 81

Earth: gifts of 31–4, 44, 67, 70, 81,
    86–7, 101, 138; as means to
    salvation 37, 52; Augustine's view of
    84–87, 95; the original ark 80
East India Trading Company 47
ecological indicators 50

economic: cycles 31
economic interests 32, 101–2; effect on
    biological diversity 32
economic objectives 4
economic progress: measurement of 50
economic structures: Reformation's
    contribution to 38
economy: Aristotle's two types 43,
    45–46; of love in Ephesians 36
ecotheological climate 33
Eddington, A. S. 56
Edwards, J.: natural events part of
    God's redemptive work 39; on the
    slave trade 42
Eichmann 132
Einstein, A. 8, 23
election of Israel 31
electron microscope 54, 64
emission credits 77
empire 11-12, 47, 76, 94, 100, 103–4,
    111, 112–14, 121, 126–7, 135
enclosure: Locke's use of 44
enemies: forgiveness of 121–2, 131
English Puritanism 38
environmental degradation: those most
    at risk from 77
EOLSS (The Encyclopedia of Life
    Support Systems) 22–3
Eusebius' account of Nicea 112–13
Evelyn, J. 29, 45
event 1-6, 12, 13, 15–65, 67–8, 70–1,
    79–83, 85–6, 97–106, 122–3,
    129–30, 137; birthing of 19–20; as
    irreducible possibility 19; seminal
    1–2, 15–27, 70–1, 79, 80–1
evolution 2, 6–7, 17, 43, 52, 59, 61–2,
    64, 66, 86, 95–6, 110, 130
excavating Jesus 123–4
Exodus story and colonization 91
expropriation 41–3, 48
externality in economics 138
extinction events 2–3, 5, 31

Fairtrade 49–50
fallout 10, 75
Father of Jesus 17, 93–4, 96, 100,
    101–2, 105, 107, 109, 114, 117–20,
    126
feminism and feminist theology 76
First historic event: the discovery of
    America 26–36, 41
food and blessing 86
forest clearance 29

forgiveness 36, 121–3, 129-39; of
  enemies 121–2, 131; in the kingdom
  of God 137–8
Form criticism 123–4
fossil fuel 11, 15, 67, 78
France 112, 131
Francis of Assisi 88–89
freedom 4, 36, 47, 59, 79–82, 99,
  100–103, 134; and labour 79, 80,
  134
freely receive, freely give 81, 86, 89, 99,
  102, 104
fresh water 8, 35, 83
Fried, E. 105
frontier expansion 29
fundamentalism 31
Funk, R.: and the Jesus Seminar 90; on
  myths Christians live by 90
Furlong, A. 118–19

Gaia 1–3, 5, 9, 20, 24, 36, 62–7, 70–2,
  80, 84, 86–7, 129–30, 134, 138–9;
  and the Apollo-Gaia Project 5; birth
  of 20; collective noun 130; and
  death 66–7, 70–1, 84–7; as Greek
  goddess 24, 36; and forgiveness
  138–9, 139; Theory 1, 9, 24, 36,
  63–4, 139
Gaian feedback systems 5, 24, 31, 132
Galileo 52–6, 64; his opponents 55–6;
  The Message of the Stars 64
Galston, D. 88–9
Galtung, J. 110–11, 126, 134–5
GDP (Gross Domestic Product) 9–10,
  49–50
Geering, L. 118–19
Genesis 1–2, 6–7, 16–21, 23, 25, 31–2,
  44, 46, 65–7, 74, 96; and dominion
  44, 96
geohistory 9, 13–14, 81
geological time 1, 2, 82
German Pietism 38
gift: in German 70, 83; its
  unconditionality 121–2; of Jesus 4,
  102–3, 98-107
gift events 3–4, 14, 33, 68, 70, 78–87,
  106; as living thread 82-5, 100;
  present givers and future receivers
  68; their ambiguity 68; their
  character 85–6
givenness 6, 17, 27, 41, 64–83, 103,
  123, 130, 133, 139; of climate
  events 67–9; of death 70; of events
  6, 17, 27, 64–79, 82–3, 130, 133;

of Gaia 6, 17, 27, 41, 64–83, 103,
  123, 130, 133, 139;
givers: relationships between 99–100
giving: dynamics of 81; for-giving 122;
  puts recipient in debt 86
global capitalism: Reformation's
  contribution to 38
GNP (Gross National Product) 9–10
God, 1–4, 7,–8, 11–18, 21, 23–6,
  29–42, 44–8, 51–2, 59, 62, 70–1,
  74–6, 79–80, 84–116, 118–127,
  129–31, 133–9; and change 1, 3–4,
  7–8, 11, 13–14, 16, 23, 30–6,
  39–40, 52, 70–1, 84–5, 88–91, 93,
  95–6, 109, 121, 123, 129, 134; of
  the creeds 109; determined to hold
  on to nothing 88–9, 99; and divine
  punishment 34–5, 46, 70–1, 84–5,
  108–9, 134; embarrassed at the
  prospect of possession 88, 99;
  feminine dimension of 36; the giver
  of images 88; gives everything away
  136; as holy 26, 72, 94–5, 114;
  images of 8, 14, 25, 31–2, 76, 91,
  93, 95, 109, 129, 130–1; imposes no
  conditions 98; of Jesus 90, 106–16,
  120–1, 134; the name of 3–4, 24–5,
  97–8, 101; as non-locatable 15–16;
  as powerful 4, 7, 16, 18, 25, 36, 46,
  88, 92–3, 101, 103, 114, 129–30; as
  suffering 7, 14–15, 30–1, 35, 84–5,
  94, 102, 105–6, 109; as transcendent
  14–16, 18, 31, 80, 92, 94–5; as
  unaccountable 15–16, 36
Goethe, J. W. v. 55
Gospel of Thomas 89, 109, 124, 133
grand narrative 1–2, 9–14, 20, 23,
  25–6, 29–30, 34, 74–5, 88–9, 96–7,
  114–15
Gutierrez, G. 127

hamartanein (miss the mark, sin) 135–6
Haraway, D. 15
hard currency and land appropriation
  45
heathens: Wesley's appreciation of 46–7
heresy charges today 118–19
Hesiod 9
Hinkelammert, F. J. 43–6, 48–52, 126,
  130
historical Jesus 36, 116-17, 123-25,
  131–2; reconstruction of 124
history, 1–3, 7, 10, 12–16, 26, 28–30,
  34–5, 47, 49, 61–5, 68–71, 74–5,

77, 82, 92, 94, 107, 109, 115–23,
    125, 130, 138; of earth 1, 10, 12,
    15–16, 30, 34–5, 49, 63–5, 68,
    70–71, 82, 107, 121, 130, 138; of
    salvation 15–16, 116
Hochma (Wisdom) 37
homo capitalisticus, homo sovieticus 81
homocentrism 63–4
hos kai (as indeed) and forgiveness 123,
    137
Hubble Telescope 58, 60, 63–4
human: alienation from earth 38;
    arrogance 71–72, 72; dependence
    on whole community of life 14–15,
    33–5, 92, 97; intelligence 25, 60,
    72, 130–31; labour 80; population
    26, 33–4, 63–4, 126; relationship
    with God 1, 4, 36, 140;
    understanding, limits of 20
Human Development Index 51
humans: as aliens 11, 24, 30; as part of
    the whole 24–5; as participants in
    the mystery of giving 37; as products
    of past and present gift events 6, 20,
    25, 28, 39–40, 66, 83–4; as shaped
    by the planet 88; as stewards of the
    planet 34–5

IHS: meanings of 116
I-It relationship 36
images of God 9, 15, 26, 32–3, 77, 92,
    94, 96, 110, 130, 131–2; destroyer
    in Revelation 94; hierarchical and
    patriarchal 95–6; imperial 48, 89;
    Paul's image of weakness 97–8;
    powerless or powerful 93–4; that
    require the death of Jesus 110
images: the power of 91
IMF (International Monetary Fund) 31
immortality 72; and the Anthropic
    Cosmological Principle 72
imperial apocalyptic Christianity 119
imperial mindset 110–12, 121
imperial power 12, 14, 26, 89, 94–5,
    105, 111–12, 114, 116, 122, 125–6,
    129
imperial takeover of Christianity 12,
    112, 113–16, 119–20
imperialism: Centre and Periphery
    nations 112–14, 129, 135–6
India: franchising of 48
indigenous peoples 32, 74–5, 78–9
Infinite universe 57, 59–60, 62
instantiation 20

intelligence 25, 60, 72, 130–31
interdependent co-arising 22
interests: economic 33, 39–40, 134
inter-human gift relationships 84
International Q Project 125
internet 9–10, 29–31; cafes 31
in-the-world asceticism 39
IPCC (Intergovernmental Panel on
    Climate Change) 9–10, 13–14, 16
Iraq body count 31–2
Israel 13, 26–7, 32, 41, 75–6, 101–2,
    107, 109–10, 114, 124, 130, 132,
    135–6; God's election of 32
It is what it is says love 106

James, W. pragmatic way of taking
    religion, 107
Jantzen, G. 138–9
Jefferson, T. on banks issuing money 52
Jesus, 3–4, 12, 14, 18, 26, 35–7, 42,
    44–5, 53, 55, 57, 82–3, 85–8,
    90–92, 94–5, 97–140; arrest of and
    Peter's denial 128; crucifixion of 110;
    death of and redemption 35–6; and
    an ecology of love 121; healing on
    the sabbath 102; his first passion the
    kingdom of God 107; his image of
    God as Father 18, 94–5, 97, 101–3,
    106, 108, 110, 115, 118–121, 127;
    his images of divine economy 37,
    107; his words and Christianity 119;
    of history 37, 117–18, 124–6, 131–2;
    and Judaism 119, 126–7; on love of
    enemies 31, 89, 94, 105–6, 109–10,
    116–23, 125, 131–2, 137; on the
    periphery not the centre of power
    113–14, 129; and Pilate 105, 107,
    109, 136; recovery of his teaching
    132; the rich young man 90, 108; the
    Samaritan woman 100; significance of
    his death 85–6, 105, 109–10, 125–6;
    then and now 107; through the lens
    of salvation 110
Job 2, 18–19, 22, 51
Jubilee Year 2000 133–4
Judaism and Jesus 119, 126–7
Julian of Eclanum against Augustine
    85–6, 96

Keller, C. 4, 12–13, 17–19, 65, 68,
    121–2, 124, 126; on novelty of
    creation 17–18, 65, 68
Kember, N. 120
Kennelly, B. 3–4, 34, 81, 89–91, 93–4,

97–99; on the mystery of giving
3–4, 34, 81, 89, 98
Kepler, J. 57
kerygma (proclamation of cross and
resurrection) 109, 110–11
khremastike (economy of acquisition)
44, 46–7
King, M. L. 52–3
kingdom: a community of subjects 102;
of gift relationships 100–101
kingdom of God 2–3, 12, 40, 60,
99–107, 110, 114–15, 122–3, 127,
137–9; according to Jesus 12,
100–105, 107, 110, 114–15, 122–3,
137; the climate of 102; the power
of the powerless 101
Klinkenborg, V. 33
knowledge building and environment
111
Koyré, A. 59–60, 62, 65
Kwok, P.-L. on discovery of America
74–7

La Cena de le Ceneri (Bruno, G.) 60
labour: as a life-purpose willed by God
42–3; as the purpose of life 41
land: appropriation and hard currency
46; possession and identity 30, 75–6,
92
language 2, 10, 12–14, 24, 29–31, 50,
60–61, 97, 130, 135–6; of
commodity relations 50; gender-
specific 29–30; of universal science
60–61
legacy effects 126
Leviticus 25, 134
liberation theology 126–8
life: more-than-human forms of 34, 130
lifestyle 4, 6, 11, 36, 44, 47–8, 52, 57,
62, 64, 69, 85–6, 131–3; choices
36, 85–6
limits of human understanding 20
Locke, J. 45–7, 49–51, 73–4, 80; and
colonial property policies 45; his
enduring legacy 49; his personal
fortune in slave trade 46–7; on
property for the accumulation of
wealth 50; on property in the state
of nature 49–51; on the right to
property 46, 49 on the 'state of
nature' 47; Two Treatises on
Government 73
London Underground bombings 744
Long, A. 66, 73

Lorenz, E. 7
love and liberation theology 126–8
love of enemies 106, 118, 120–21, 122;
and ethnic identity 119
love of the world 47, 138–9
Lovelock, J. 1, 2, 10, 21–2, 25, 35–6,
37, 64–7, 81–2, 131; on the genesis
of Gaia 66–7

Manhattan Project 61–2
manifest destiny 32
Mann, T. 121–2
Margulis, L. 22–3, 65; on humans'
place in nature 65
Marion, J. L. 70, 82–3, 86, 95–6, 101;
on gift 101; on the context for
giving 82–3
Marxism 127–8
mass production and individual lives 43
mathematics 61–3
metalanguage 10
metanoein (change of mind) 136–7
Methodism 47
microscope 55, 65
military power 36, 129
mineral deposits 30–31
mobile phones 31
modern culture: influenced by
Reformation 39
monetary growth 51
monetary wealth and alienation 44
money: in circulation 51–2; giving a
right to greater possessions 46;
origins in Temple tributes 44–5; as
product 46–7; use of in exchange for
land 46–7
more-than-human 34, 130
Moses and the thirsty kid 101–2
music 20–21
mutually assured vulnerability 4, 5
mystery of giving 3–4, 15, 34–5, 37,
81–2, 84–9, 98, 100, 101, 103,
106–7, 129; Christian expression of
15, 35, 37; and conventional
economic analyses 86–7;
encompasses life and death 84–6,
100, 101; in a new theological
climate 98
myth of identity 75

Nag Hammadi 124–5
name of God 3–4, 25–6, 98–9, 102
narrative: grand 1–2, 10–15, 21, 24,
26–7, 30–31, 35, 75–6, 89–90,

97–8, 115–16; scientific 10–11, 14, 24
natality 19, 25, 65, 77–8, 138
native species 32
natural life support systems 35, 50–51, 64
Nature 1–4, 6–7, 11–12, 14, 15–18, 20, 23–4, 27–8, 35, 39, 47, 49–51, 55, 57–59, 61–67, 71–3, 76–8, 81, 84-6, 90, 87–101, 104–6, 120, 116, 119, 126, 128, 131–3, 137–9; Locke's views on 48
Nature's gifts: possession of for profit 90
Nazism 43, 53
Nelson-Pallmeyer, J. on violence-of-God traditions 9, 32–3, 92, 95
New England forests 46
Nicene creed 112–16; as criterion for admission or rejection 113; enforcing dominance of bishops and emperor 113–14; and the filioque clause 115; and potential for Christian violence 114
Nicholas of Cusa 57
Noah 32, 91–2, 97, 110
nonviolence 12, 37, 111–12
nonviolent theological norms 15, 35, 111
Northern Ireland 42, 119–20
novelty in creation 17–18, 65, 68
nuclear bomb 61–2
nuclear energy 79
Nuremberg trials 132

O Crualaoich, G. 72
observership: non-geocentric 60–64
O'Connor, F. 81
oikonomike (economy of basic needs) 44, 46–7
omniscience of God 16–17, 63–4
Oppenheimer, R. J. 61–2
option for the poor 128
organism 2, 4, 10, 21–23, 25, 29, 34–5, 55, 64–5, 67, 83, 102
origin of the universe 21
origins 6, 19, 21, 43-5, 71, 77, 82, 115–17, 126
orthodoxy 28, 127–8; defence against liberation theologians 127–8
other-worldly asceticism 39

parable: the Last Judgment 83; the Samaritan 120–21

patriarchal images of God 95–6
patriarchy 26, 76–7
Paul 13, 83–5, 87, 97–9, 109–11, 125–26; his kerygma 109–11; on givenness 83–4; on the weakness of God 97–8
Pelagius 96
Pentateuch 26
physics 22, 29, 60–63, 67
Pietism: German 39
Pilgrim Fathers, 75–6
political structures: Reformation's contribution to 39
population 26, 30–31, 33–4, 49, 63–4, 70, 126
postcolonial 13, 31, 42, 51, 74–77, 121–2, 124, 126, 131–2; domination, 31
postmodern theology of the cross 122
power: and knowledge in an imperial mindset 111–12; to forgive 136–7; imperial, 12, 14, 26, 89, 94–5, 105, 111–14, 116, 122, 125–6, 129; over 19, 26, 37, 89, 94, 102, 115, 130, 131; within the kingdom of the powerless 137
Prigogine, I. 8, 58–9
production 11–12, 32–3, 41, 43–4, 50–51, 60, 124
progress 10–11, 18, 31, 51
property 43–52, 76, 102–3; destroyed in its appropriation 43–4; development of 43; ownership 44, 49–50; in the state of nature 49–51; wealth accumulation 42, 49–51
Protestant asceticism 41–2; Reformation 13–14, 28, 38–39, 39, 42, 43–44, 53–54, 55, 119–20, 124
Puritanism: and spontaneous enjoyment of life, 41–2; writings on the security of possessions 40–41

Q people 127–9
Q Sayings Gospel 37, 110, 125, 128

radical change 56–7, 63–4
Ratzinger, J. and liberation theologies 126–7
receivers and givers 4, 22, 67–8, 70, 82–3, 100–101
recontextualization, a challenge to theologians 24
Redaction criticism 124–5

Reformation: Protestant 13–14, 28, 38–39, 42–4, 53–5, 119–20, 124; shaping political and economic structures, 39

relationships: between Earth, humans and God, 24, 37; 'real' 36
religious imperialism: resources for dealing with 124
religious language 12, 130, 135–6; meanings 3–4, 12, 16, 24, 39, 42–3, 57, 93–4, 102–3, 109–11, 116, 120–21, 135–6; re-envisioning narratives 34; shifts in perspectives 35; utilitarianism 36; violence 32–3
repentance 40, 123, 136–7
Rilke, R. M. 92–3
Robinson, J. M. R. 37, 108–9, 119, 124–8, 134–6
Roman persecution of Christians 95
Rome and Constantinople 89–90, 95–6, 105, 109, 114–15
Rosenzweig, F. 17–20
Royal Society 29–30
Rudwick, M. 10

sacred 2–3, 4, 9, 15–16, 32–3, 92, 110–11, 139
Salt Tax 52–3
salvation 15–16, 32, 35–6, 38–9, 53, 60, 87–8, 103–4, 110, 114, 116
Sayings Gospel Q 37, 125
Schrödinger, E. 22–3, 29, 57–8
Schwartz, R. 29, 75–6
Schweitzer, D. 40
science 1–3, 5–12, 13, 21–5, 28–30, 32, 53–64, 72, 82, 126; discourse in 10, 24; male-domination of 29–10
Scripture: effects of translation 38–9
Second historic event: the Protestant Reformation 37–54
Second World War 132
Segundo, J. L. 127–8
self: interest 11, 33, 42, 46; making, 34; perception, 15; regulation, 23; understanding, 35
seminal event 1–2, 16–28, 71–2, 80, 81–2
sexual violence 75
shrinkage of the globe 28–9
sin 35–6, 47, 71–2, 85–8, 109–10, 114, 116, 128, 135–7
situatedness of creation 20
slave trade and exemplary Christians 43

Snyder, G. 88
society, symbolic order and social order 78
soil contamination 12
Sophia (Wisdom) 37
Source criticism 124–5
South Africa 123, 132
Spivak, G. 121, 126
Stăniloae, D. on human labour 80–81, 135
Stengers, I. 7, 58–59
sub-prime mortgages 49–50
suffering 8, 14–16, 31–2, 34, 36, 72, 85–6, 95, 103, 106–7, 110
Sufi 8
survival 33–5, 40, 64, 135
Suzuki, D. 24–5
symbiotic givenness 68

Talmud 8
technology 21, 23, 29, 31, 61, 90
tehom (the deep in Genesis) 1–2, 6–7, 18, 19
telescope 28, 51–6, 58, 60, 63–4
Temple tributes the origin of money 44–5
terra nullius (ownerless land) 46
terrorists 74, 77
texts: exegetical layering in 124–5; stratigraphic anaysis of 124–5
theological change 130
theological climate built on the mystery of giving 98
theological relationships 35
theological shift in focus 122
theology 1–4, 6, 12–15, 18, 26–7, 31, 37, 47, 62, 77, 79–81, 89, 92–3, 96–7, 103–10, 116–23, 125–9, 131–32, 135, 137–9; of gift events 80;
thinking: analogical-rational, 27; rational-instrumental, 27, 60
Third historic event: the invention of the telescope 53–65
thread of life 84–5
Clotho, Lachesis and Atropos 84–5
time: geological 1, 2, 83
Tipler, F. 62–3, 72
tipping points 84
tohu (formlessness in Genesis) 1–2, 6–7, 18, 20–21
toxic waste 9
Tradition criticism 124–5

traditions, violence-of-God 9, 15, 31–3, 92, 95, 97
Truth and Reconciliation Commission 132
Tudge, C. on our debt to trees, 87
Twain, M. 70

UN Climate Conference 2007 78
UNCED 'Agenda 21' 78
UNEP 9–10, 13, 14–15
UNESCO 23
United States 36, 40, 49–50, 52
universe 18, 21, 24–5, 27, 55, 57–64, 66, 71–4, 85, 131; as infinite 57, 59–62
urbanization 31, 50
US Constitution 30
utilitarian view of earth 15, 38

vacuum domicillium (waste land) 46
violence: in biblical texts 92; religious 32–3; violence-of-God traditions 9, 15, 31–3, 91, 95, 97
violent images of God: in Christian worship 92; their origins 94–5
virtual earth models 69
Volk, T. 67–8

war against terrorism 77, 133
Wasdell, D. 5
waste 5–6, 9, 32, 45–6, 50–51, 70, 81, 91, 134, 139
water 17–19, 21–22, 26, 28–9, 67; fresh 9, 36, 84; privatization 52–53
WCC (World Council of Churches) 121–2
wealth 30–31, 40–53, 63–4, 89–90;

accumulation 43–4, 49–51; ethics of 41
Weber, M. 15–16, 32, 39–45, 51; on the formation of modern culture 39; on the origins of capitalism 43
Weeden, T. J. 108
Weil, S. 28, 31, 43, 95, 105, 114; on Hitlerism 43; on justice and the use of force 105; on totalitarianism of the church 114
welfare and economic growth 33–4
Wesley, J.: arguments against slavery 48; imperial policy the cause of human misery 49; on pre-eminence of the heathen, 47–8; on pursuit of profit, 42; vision of 'the community of goods' 47–8
West, C. 127
Western consciousness 12
Whitehead, A. N. 53–5
Wisdom 1, 37, 66, 72–4, 87, 98–9, 108, 115, 136; of Solomon 66, 73
Wittgenstein, L. 24
womb 19, 20–21
women 36, 77–9
work: for the benefit of the human soul 38; a life-purpose willed by God 42–3; and profit 38, 41, 81; as proof of genuine faith 42
workers: in commodity markets 50
world alienation 43–4
World Bank 31
world marketplace 76

Zeus 94–5

Related titles from Routledge

Gaia's Gift: Earth, Ourselves and God After Copernicus
Anne Primavesi

'Turning traditional theology upside down, Primavesi offers an earth-centred vision that is liberating and profoundly hopeful. Written with eloquence and clarity, *Gaia's Gift* is the most up-to-date theological response to the challenge of science. A tour de force!'
Kwok Pui-lan, Episcopal Divinity School, Massachusetts, USA

'Once again this unique and important thinker breaks new ground. Primavesi brings her distinctive blend of critical acumen, cross-disciplinary savvy and ethical passion to bear upon the interface of cutting-edge theology and radical science.'
Gwen Griffith-Dickson, Gresham College and Birckbeck College, University of London, UK

Gaia's Gift, the second of **Anne Primavesi**'s explorations of human relationships with the earth, asks that we complete the ideological revolution set in motion by Copernicus and Darwin concerning human importance. They challenged the notion of our God-given centrality within the universe and within earth's evolutionary history, yet as our continuing exploitation of earth's resources and species demonstrates, we remain wedded to the theological assumption that these exist for our sole use and benefit. Now James Lovelock's scientific understanding of the existential reality of Gaia's gift of life again raises the question of our proper place within the universe.

ISBN13: 978-0-415-28834-7 (hbk)
ISBN13: 978-0-415-28835-4 (pbk)

Available at all good bookshops
For ordering and further information please visit:
www.routledge.com